DISTINCTION
JOURNAL OF FORM
Volume 1 Issue 1

Spring 2025

The Spencer-Brown Society

Editors:
DIVYAMAAN SAHOO
FRED CUMMINS
LEON CONRAD
GRAHAM ELLSBURY
FLORIAN GROTE
RANDOLPH DIBLE
PHILIP FRANSES

Copyright © 2025
The Spencer-Brown Society,
and the authors.
All rights reserved.

ISBN: 978-1-84890-502-3

Published by College Publications, London.
Spencer-Brown Society Logo © Akiem Helmling

Announcement of a New Journal

The First *Distinction*

Distinction: Journal of Form is the journal of The Spencer-Brown Society. In 1969, mathematician and polymath George Spencer-Brown (1923–2016) published *Laws of Form,* a revolutionary work in pure mathematics in which the central concept is the simple act of drawing a distinction. From this follow two laws and these are the laws of form. The intuition that the act of distinction lies at the foundation of everything characterizes *Laws of Form* and makes the work relevant beyond the discipline of formal logic. It is from this innovative concept of distinction that the Society and this journal draw their inspiration.

Cybernetics, semiotics, and systems theory have incorporated key insights from *Laws of Form,* which has influenced, for example, Francisco Varela and Humberto Maturana's theory of autopoiesis and Niklas Luhmann's social systems theory. Spencer-Brown led the spiritual interpretation of his work in *Only Two Can Play This Game* (1971) and in *A Lion's Teeth* (1995), in which he refers to *Laws of Form* as the vehicle that he used for his enlightenment and that others can use for theirs (1995, 148).

Distinction: Journal of Form aims to foster diverse interests in new ways and carry Spencer-Brown's revelations further than previously. To this end, in this journal, the Society intends to publish the proceedings of the ongoing *Laws of Form* conference series, related conferences, such as the 2023 *Thinking the Float Tank* conference, future workshops, and other relevant works that fall under the scope of this mission.

For other publications and related works, the Spencer-Brown Society provides the book series *Marked States: Series on Form.* Both the journal and the book series are published by College Publications Ltd.

Randolph Dible
Communications Director
The Spencer-Brown Society

Editorial Introduction

Welcome to the inaugural issue of *Distinction: Journal of Form*.

In August 2019, in the charming oak-panelled Old Library, School of the Arts, University of Liverpool, a conference, LoF50, was held to celebrate the 50th anniversary of the first publication of Laws of Form, George Spencer-Brown's pioneering work in the foundations of mathematics, of which Bertrand Russell said: "In this book G. Spencer Brown has succeeded in doing what, in mathematics, is very rare indeed. He has revealed a new calculus, of great power and simplicity."

The papers presented at LoF50, which were diverse in nature, were subsequently published in *Laws of Form, A Fiftieth Anniversary,* (World Scientific Publishing, 2023), a volume of over 900 pages.

At the close of LoF50, the attendees voted unanimously to hold a second conference, two years hence, in 2021. Unfortunately, due to the global Covid epidemic, that conference had to be postponed until the following year.

Thus, in August 2022, LoF22 was held, like LoF50, three years previously, in the Old Library at Liverpool University. The University generously provided excellent audio-visual facilities to enable those presenters who, unable to travel to Liverpool due to continuing Covid restrictions, were nevertheless able to make their presentations online.

The Spencer-Brown Society

In 2023, to honour the memory of George Spencer-Brown, The Spencer-Brown Society was formed to promote and advance his work. For more information about the Society, please refer to the Society's constitution which can be found at the rear of this volume. A committee comprising: Graham Ellsbury, Chair; Florian Grote, Secretary; Leon Conrad, Treasurer; Divyamaan Sahoo, Publications Director; and

Randolph Dible, Communications Director, was elected at the inaugural meeting in 2023, and re-elected at the AGM in August 2024. Organising future conferences and arranging publications are among the principal tasks of the committee.

Distinction: Journal of Form

It is in this respect that the Society decided to launch *Distinction: Journal of Form*, to publish the proceedings of the LoF conferences and other works concerning and related to *Laws of Form* and Spencer-Brown's mathematical and other writings. College Publications was selected as the publisher since, unlike many academic publishers, the authors and contributors retain the copyright to their work.

Laws of Form has been, and continues to be, highly influential in a number of fields, including, the foundations of mathematics, philosophy, cybernetics and social systems. This Journal provides a channel for the publication of papers authored by the diverse group of researchers who draw heavily on Spencer-Brown's work, and a resource for those who wish to explore recent thinking and perspectives in this domain. Consequently, papers in this and future issues of *Distinction: Journal of Form*, will reflect the interdisciplinary nature of research in these fields and will be as diverse as those of *Laws of Form, A Fiftieth Anniversary*.

To ensure that the papers presented at LoF22 would be published in a more manageable form than those of LoF50, it was decided that they should be published, not in a single, but in two, issues. This current issue is the first fruit of these decisions and therefore represents the first of two issues of proceedings of LoF22. The second issue will follow in due course.

Future publications

The Society's 2024 conference, LoF24, like its predecessor conferences, LoF50 and LoF22, was held in the Old Library at Liverpool University. The proceedings of LoF24 will be published in *Distinction: Journal of Form* in due course.

LoF50, LoF22, and LoF24, were arranged in conjunction with West Den Haag, the international platform for contemporary art, which professionally video-recorded all three conferences. In August 2023, West Den Haag, at their gallery, the former U.S. embassy, hosted *Thinking the Float Tank,* a conference to celebrate the fiftieth anniversary of the AUM conference held at the Esalen Institute, in Big Sur, California, in 1973.

Future conferences and publications

The Society intends to hold Laws of Form conferences every other year. The next conference, LoF26 will be held at the University of Cambridge. Other related conferences, outreach events, and publications, including a book series, *Marked States: Series on Form,* are in the pipeline, and will be announced in due course.

Funding conferences

We understand that travelling to conferences, especially for those who travel from abroad, and obtaining accommodation, represents a significant cost for many members.

To avoid placing an additional financial burden on attendees, the committee is firmly opposed to introducing membership fees for the Society, or fees for attending, or presenting at, the Society's conferences.

We would therefore prefer that the Society and its conferences are funded by voluntary donations. However, we will have no option other than to introduce fees if we are unable to attract sufficient funding.

If you are in a position to be able to donate to the Society, we would hugely appreciate any donation you are able to give.

Donations may be made here:

$$\text{https://lof50.com/}$$

I wish to give heartfelt thanks to those who have already donated, some most generously, to the Society.

If you would like to become a member of the Society, please simply fill in the form on our website.

Acknowledgments

Many individuals have contributed time and effort towards making the conferences a success. Not all can be listed here, but the Society particularly wishes to thank:

Andrew Crompton, for arranging with Liverpool University, the use of the Old Library for the LoF50 and LoF22 conferences; Mike Zundel for arranging the use of the Old Library for LoF24; and the University of Liverpool for generously hosting the conferences and providing comprehensive audio-visual facilities;

Steve Watson and the Faculty of Education at the University of Cambridge, for offering to host the forthcoming Laws of Form conference, LoF26;

West Den Haag for their cooperation with all three conferences, professionally video recording the presentations, and hosting *Thinking the Float Tank;* Akiem Helmling, for making his GSB notation typesetter available online, producing the final design for Spencer-Brown's gravestone, and creating the Society logo; Lou Kauffman for arranging the publication of *Laws of Form: A Fiftieth Anniversary* with World Scientific Publishing; and Kevin German for providing his React Library HTML image generator.

The Society would also like to thank the many reviewers for their time and diligence in peer-reviewing the papers in this volume.

As Chair, I particularly wish to extend my personal and heartfelt thanks to the members of the committee who are working hard, and giving freely and generously of their time, to produce high quality conferences and publications:

Florian Grote, for conference organisation, taking on the role of Secretary of the Society, administering the Society, providing and administering the peer-review software system, numerous communications with the authors, and administering the website;

Leon Conrad, for conference organisation, taking on the role of Treasurer of the Society, efficiently managing and reporting on the Society's finances, editing this volume, and administering the website;

Divyamaan Sahoo, ably assisted by Fred Cummins and Philip Franses, for editing this volume, and meeting the many exacting challenges posed in typesetting this Journal, including the transcription into LaTeX from source papers and from the video recording of Stephen Wolfram's keynote talk. This volume could not have been published without their energetic commitment;

Randolph Dible, for publicising the conferences, administering the Society's communications, coordinating arrangements with College Publications, acting as Series Editor for *Marked States: Series on Form*, and conceiving and directing *Thinking the Float Tank*.

The committee is giving serious consideration to appointing an advisory board to assist and support the committee in its decision-making and an editorial board to assist with publications. If you would like to contribute to either board, please contact the committee outlining what you feel you can contribute. We are also considering incorporating regular features in future editions of *Distinction*, including a column, and a book review section. If you would like to contribute to either, please contact us.

Finally, with its unmatched scientific heritage, there is no more appropriate location for the Laws of Form conferences than Spencer-Brown's alma mater, the University of Cambridge. We are therefore delighted to announce that LoF26 will be held throughout the week of 10th–14th August 2026, at the Faculty of Education, the University of Cambridge.

We look forward to welcoming you to LoF26 in Cambridge in August 2026!

Graham Ellsbury
Chair
The Spencer-Brown Society
https://lof50.com/

INVITED KEYNOTE: STEPHEN WOLFRAM

STEPHEN WOLFRAM
Wolfram Research
s.wolfram@wolfram.com

This is an edited transcript of a keynote presentation made at LoF22. Images are excerpts from a wealth of information presented visually. The talk can be viewed at https://youtu.be/AXFVmJHFeX0?si=Zse_vg7kcPpitz0P

STEPHEN WOLFRAM:
I realize that I've had this book [*Laws of Form*] on my shelf for a very long time, and I ended up talking about it a bit in the book that was just shown, which came out in 2002 [*A New Kind Of Science*, 2002]. On one page of this book, I talked about having found the minimal axiom system for Boolean algebra. And, as part of the notes for that, I talked about *Laws of Form*.

I was searching my archives to see what kinds of interactions I'd had with George Spencer-Brown, and I have to tell you of the introduction that I had to him. This was in December of 2003. A mail message was sent to me from a customer service person at our UK office, which reads as follows: "I have just been speaking to Professor Spencer-Brown from Warminster. He said he received a copy of NKS [*A New Kind of Science*] from Mr. Wolfram and wishes to speak to him regarding it. Whilst talking to him on the phone, he informed me that he was dismissed from the RAF for flying his plane into the Blackpool Tower during a fly-past for the people on Blackpool beach. I said that it's good he's still here to tell the tale, and he told me that, in fact, he's written a book about flying and was only able to do this because he had so much free time after having been dismissed." Anyway, the message speaks of an interesting personality. I ended up talking to George Spencer-Brown on the phone, and we ended up having several interactions and getting him, I think, a computer and a copy of Mathematica.

What I want to talk about here is getting to the foundations of the foundations of things, which I think is a topic of interest in the spirit of *Laws of Form* and of George Spencer-Brown, and talking about just how simple things can be underneath, and what those simple things can produce. I should comment on the one line in *A New Kind of Science* that is a reflection of my efforts in Boolean algebra. The question that I was interested in answering was: if you just consider possible axiom systems in mathematics as represented by equations, for example, if you just start enumerating them, where do the ones that [ground mathematics] show up?

So the question is, if you imagine that you're generating possible equations or axioms at random, for example, how far do you have to go before you reach axioms that correspond to things that we've studied in mathematics? And so the thing that I went to do was to find "what's the very simplest possible axiom system for Boolean algebra?" And that's the answer there (Eqn. 1), where we can interpret dot here as a NAND operation or a NOR operation.

$$((p \cdot q) \cdot r) \cdot (p \cdot ((p \cdot r) \cdot p)) = r \tag{1}$$

Interestingly, the proof of the correctness of that is something you can now generate in the Wolfram Language in a couple of seconds. But an interesting feature of it is that in the 22 years since I first derived this, or at least my computer first derived this for me, nobody has really been able to say much that is of human value about what the proof of this means. And it's an interesting example of raw computation meeting things of mathematical interest.

Maybe I should describe a little bit about the journey that I've been on of trying to understand the foundations of the foundations of things. I was lucky enough when I was a kid in England in the 1970s to do a bunch of work on particle physics and cosmology and things like that. And so I got involved in trying to understand the foundational questions about the universe in terms of what one might call traditional physics. Back then, [in order] to do the kinds of computations needed in physics, I started developing computer systems and developed the thing called SMP—Symbolic Manipulation Program—which was a forerunner of Mathematica and Wolfram Language. And when I started developing that in 1979, one of the questions that I wanted to answer was: What's a good underlying representation for the kinds of computations that, as I would say it now, we humans want to do? And, you know, we had all kinds of things where you would describe variables and make loops and so on, in the traditional programming languages that hewed close to the actual hardware construction of computers. But I was interested in the more theoretical question of what would be a good model for the way computation gets done, and how we think that computation gets done, and what I ended up with was an approach based on transformation rules on symbolic expressions. And so the basis of SMP was that everything that you define was a transformation for a pattern for symbolic expression.

So, you know, a function would be, in modern notation and Wolfram Language, "f of x blank colon equals x squared" or something. And that means, if you see "f of blank" anything named "x" then transform it to "x squared." So the idea was to take symbolic expressions that were tree-structured, things with arbitrary symbols in them, and to have the whole process of computation be one that involves successive replacement of those things with transformations, which have been defined for patterns on those symbolic expressions.

The big surprise is that this approach works spectacularly well. And now it's 40 something years later, and the foundation for Wolfram Language and Wolfram Alpha, and everything we've done is this idea of representing everything in terms of symbolic expressions and having the actual core computational operation be successive rewriting of those of those symbolic expressions using patterns that represent classes of those expressions. So in a sense, it's a very simple idea about how one should think about computation that we humans want to do. And it's worked spectacularly well.

So having built SMP, I then got interested in some basic science questions, and the particular question I was interested in was: We see all kinds of complicated stuff in the world; Where does it really come from? What's the foundation for making complicated stuff? And I tried using methods from mathematical physics and those kinds of things to try and answer that question, and I got basically nowhere with it.

Then I started asking the question: Well, if we just imagine that we are making models where we're thinking about what the meta model is for how things work in the world, maybe the right meta model is more like what we see in computation and programs than what we see in mathematics, in traditional mathematics. So that got me interested in the question: if nature operates according to programs, what kinds of programs might they be? And in particular, if we imagine that those programs are simple, well, then there's a very basic science question: what do simple programs typically do? And we didn't know the answer to that.

And so I started exploring this question: What do simple programs typically do? My favourite class of such programs is cellular automata. So if I just make an example here of a very simple cellular automaton (Fig. 1), we just have a rule. In a sense this is substitution rules all over again, because what's happening is, we have a line of cells, each either black or white, and at each step, the colour of a cell is updated according to the colour of that cell before and the colour of its two nearest neighbours.

So here's a simple example of that (Fig. 1). If we say, let's actually run that rule starting from just a single white and single black cell; let's run it for 40 steps or something; let's just show it with mesh lines here. That's the result we get:

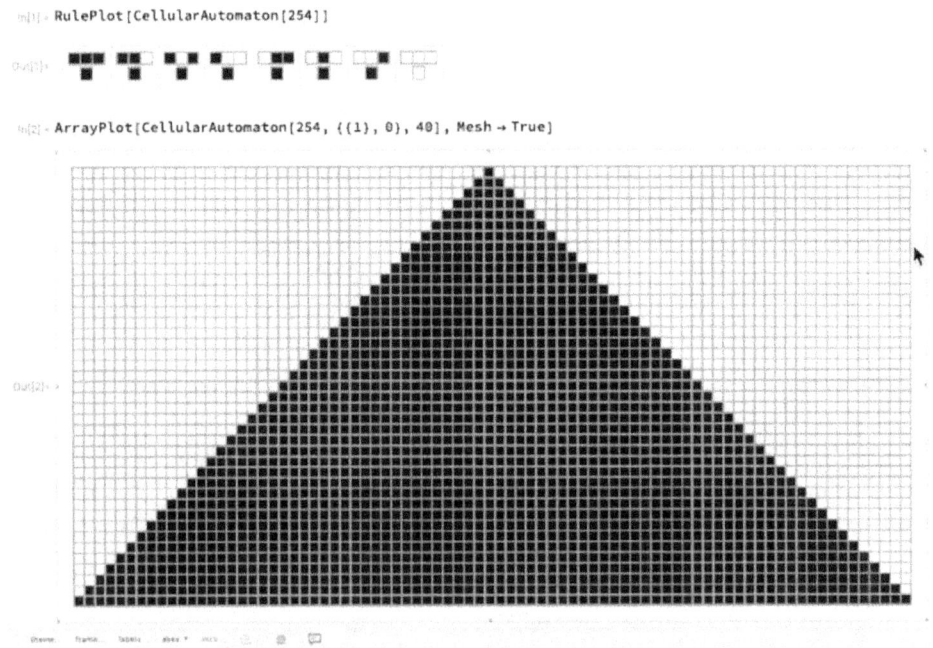

Figure 1: Simple cellular automaton.

So, very simple rule, very simple behaviour. So, let's change the rule a bit. Let's say we use this rule here [Wolfram Rule 90]. Well, then we can run this again. We'll see we get somewhat more complicated behaviour. But if we run it for a while, we'll discover that though it is very intricate, it is ultimately, in a sense, a very regular pattern. It's just a nested pattern here (Fig. 2).

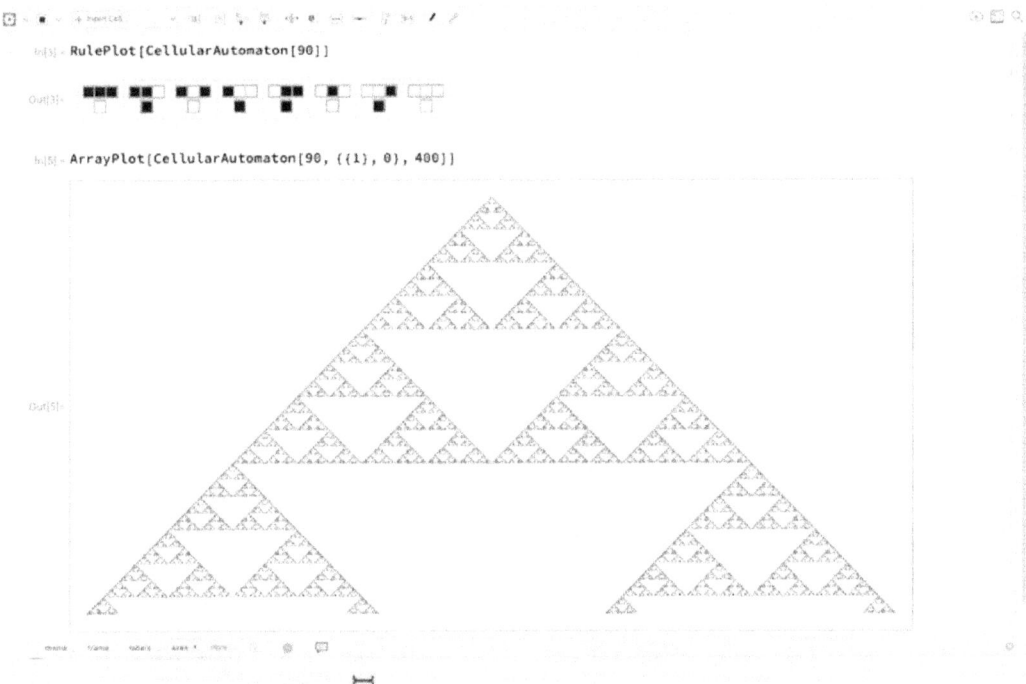

Figure 2: A variant on the automaton of Fig. 1 using Wolfram Rule 90.

So now let's imagine that we have this kind of computational universe of possible programs out there. Let's see what happens if we turn our computational telescope out into that computational universe and just find out what's out there. What possible things can happen with programs like this? In these days, it's really straightforward to do that. So we can just say, just show us, for all of these, do 40 steps of each one and let's say here we want to show the first, say, 64 of those of those rules (Fig. 3).

So each one of these pictures (Fig. 3) is a different rule of similar type, starting from just a single black cell. And we see many of those rules produce very simple behaviour. Sometimes we'll have a nested pattern. Here's one of those (e.g. Fig. 2). We keep going. We're sort of looking out into this computational universe with our computational telescope. We get to rule number 30. And we find something, that to me, at least, was extremely surprising. In fact, when I first saw this, it took me a couple of years before I really realized what I had seen. But let's try running this again up here. Let's look at what Rule 30 is. It's just a substitution rule, essentially, for these triples of cells. Let's run Rule 30 for 400 steps, let's say. Here's the result (Fig. 4):

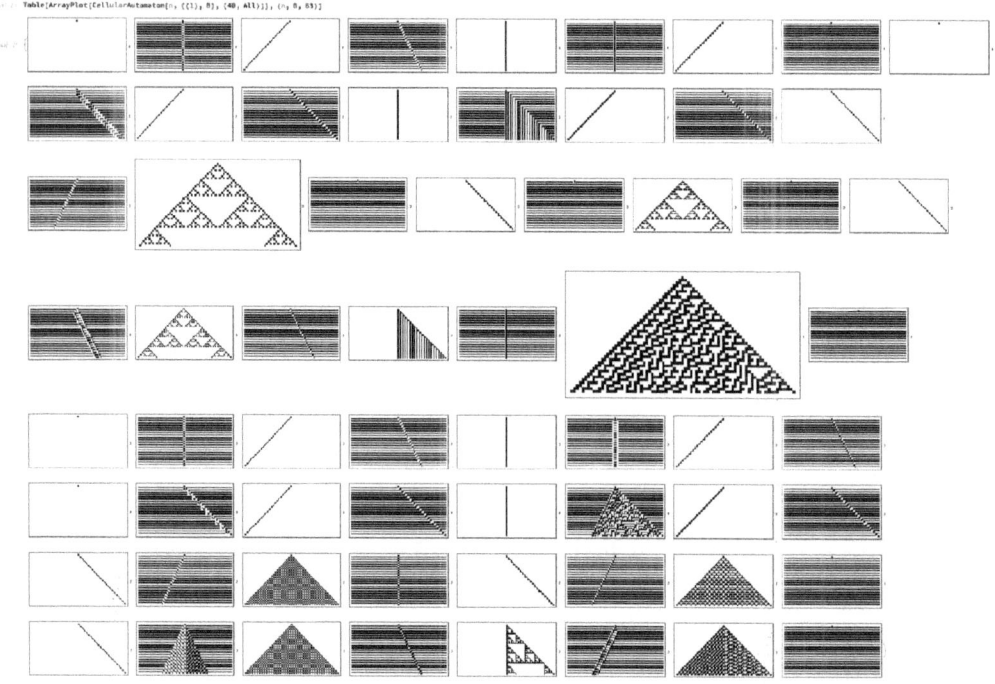

Figure 3: Patterns produced by each of 40 rules, employed in succession. Pattern produced by Rule 30 (See Fig. 4) is enlarged.

So you know what I at first thought was: That's such a simple rule. When we run it, the thing we get out must somehow be simple. So, you apply various mathematical methods, and statistical methods, and cryptographic methods, and so on, and you try and crack this. And what I found was that I couldn't crack it.

And in fact, a number of mathematicians—rather distinguished mathematicians who happened to be interested in what I was doing—also tried to crack it, and they didn't succeed. And their main conclusion was, well, then we just have to give up. What I realized—and this was probably my main achievement in those days—was that the very fact that it was hard to crack what was going on in this case was itself a significant fact. If we imagine that we asked the question: What will this thing do after a very large number of steps? To figure that out, we can just do the computation that this system itself does. But the question we might ask is: Is there some way to jump ahead? To have a smarter computation that, much more rapidly than this system itself, can figure out what it's going to do? It can just say:

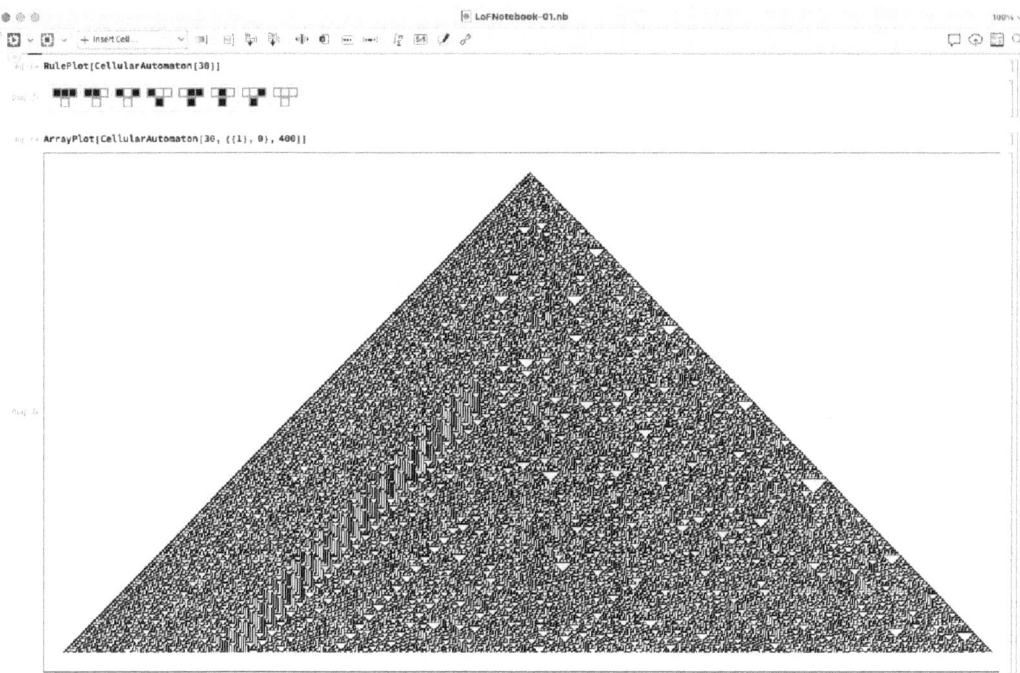

Figure 4: Pattern produced by iteration of Rule 30.

...and the answer is 42 or something, and jump to the end. So let's say we want to find what this does after a trillion steps. We want to find the centre column of this rule after a trillion steps. Can we do that much more efficiently than just running a trillion steps in this evolution?

So the surprising conclusion is we probably can't. In fact I think this is an example of computational irreducibility—an example of a case where there's really no way to find out what it does much more efficiently than just by running it and seeing what happens. That's a corollary of a thing I call the principle of computational equivalence, which is the statement that if you think about a system like this computationally and you ask how sophisticated is the computation that it's doing, as soon as you get above a very low threshold, most systems that you just pluck out of the computational universe will be equivalent in their computational sophistication. None of them will be able to jump ahead of another. So it means that if we with our brains are trying to figure out what the system is going to do, our brains are stuck at the same level of computational sophistication as this system. And that's why this looks complex to us, and why we can't expect to jump ahead of it. So the big thing from this was realizing that, yes, you could even with these very simple rules, get very complicated behaviour.

You could have certain hypotheses about things like the principle of computational equivalence which gave one predictions about what one should see in the world, like that one should typically see that, as soon as one sees complex behaviour, one will see things which are as computationally sophisticated as anything, among other things, things which are computationally universal. This is one example of a piece of evidence for this, that was from 2007, where I had found the simplest possible Turing machine (Fig. 5). This is a Turing machine with two states and three colours, the simplest possible Turing machine that does not have obviously simple behaviour. And it was a young chap who, rather quickly, managed to prove that this machine is actually universal, giving one a piece of evidence for this principle of computational equivalence. And the only reason that we don't know that more obviously, is that in doing engineering and so on, we've always tended to avoid systems where we can't foresee what they're going to do. So we've always tended to use computationally reducible systems so that we can foresee that they're going to do the things that we want to do from an engineering point of view.

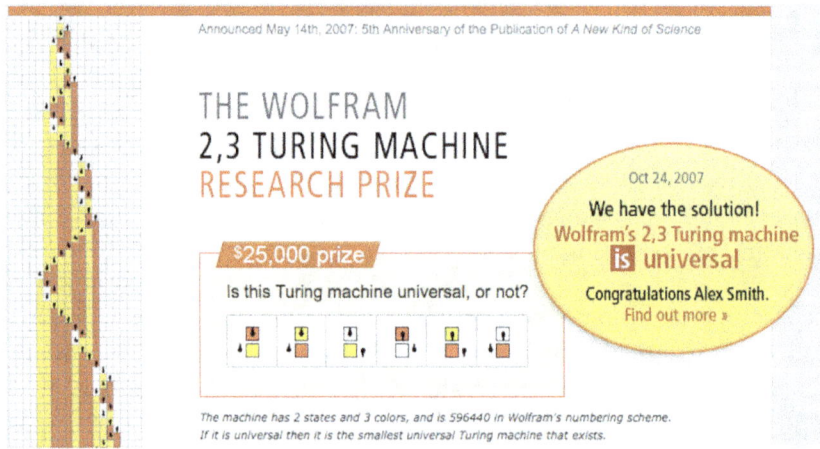

Figure 5: The smallest universal Turing machine possible.

One of the things that happened after discovering the phenomenon in Rule 30 and computational irreducibility—I went searching for just how general this phenomenon is and how relevant it is to the secret that nature has by which it makes complexity all over the place. And I came to the conclusion that this really is the core secret that nature has that allows it to make complexity; this phenomenon that in the computational universe there is this computational equivalence, and it's very easy to have systems that are capable of complex behaviour.

My big book *A New Kind of Science* is about studying the computational universe and its implications for existing sciences and so on. One of the things that came up in that book was a "use" case. The "use" case is: What about our whole universe? What about the physical world? What about fundamental physics? Is that something where these ideas of simple computational rules can be applied? I made some progress on that in the book, and then I left it for many years and pursued more the questions about building technology around ideas that came out of the science and things like Wolfram Alpha and so on. But, about three years ago now, as a result of a minor technical breakthrough, I came back to this question of: Well, What about physics? Is there a way to see whether physics is computational all the way down? And the huge surprise is that, well, yes, there is, and we kind of figured out how it works, which I consider to be very exciting. And maybe many of you have heard about this already, but I'll give you sort of an outline and then tell you some of the newer things that we've figured out. This is a visual summary of our physics project (Fig. 6).

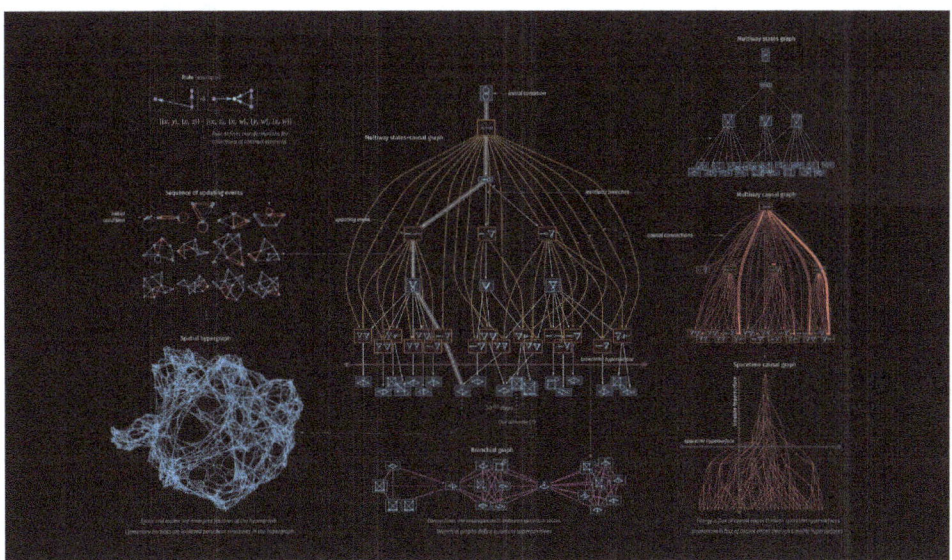

Figure 6: Wolfram Physics Project.

The real question is: So what is the universe really made of? And the first thing that we start talking about is, well, what is space made of? We would normally think from Euclid on, that space is just something where you put things in it. It's not something that's made of anything. It's just like we might say we've got a fluid like water, and we pour it, and it does all those kinds of continuous fluid-like things. And we just think of it as a fluid. But actually, in the case of the fluid, we know it's made of molecules down there. Well, the first point of departure for the physics project is to realize that space is actually not just a thing you put stuff in. It's something that itself has a structure, and the structure that is the easiest way to describe the structure is a hypergraph.

But fundamentally, space, we imagine, consists of a bunch of discrete elements. We can think of them as discrete points. The only thing we know about those points is how they're related to other points. We don't know, for example, how they're embedded in three-dimensional space. We don't know what coordinates they have. They don't have any of those things. All they are are a bunch of discrete elements. And what we know about them is how they're related to other elements. So let's say we have a simple graph here (Figure 7) that's just showing the atoms of space and showing the relation between atoms of space.

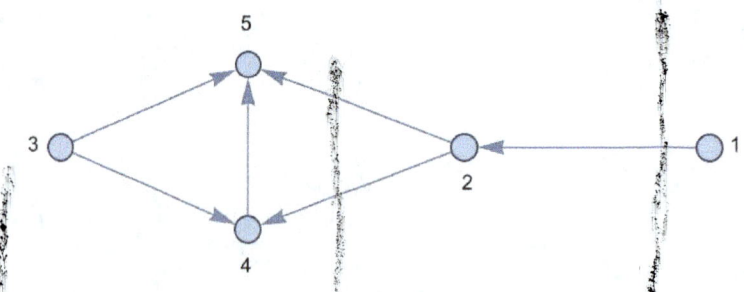

Figure 7: Graph showing relations between notional atoms of space.

Now what we imagine is that—a bit like in a cellular automaton—we imagine that there is an evolution rule for these hypergraphs. This particular one is just a graph. But we could imagine a hypergraph where instead of just an edge connecting two atoms of space, it relates more than two atoms of space. But we can just say: What is the dynamics of this? Well, the idea is that you're updating this structure, just like you're updating the line of cells in a cellular automaton. You're updating it by saying: Whenever we have a particular piece of network, let's say a piece of network that looks like this, replace it by a piece of network that looks like that. Okay. So let's try doing that. So in this particular case, we start off from that initial condition.

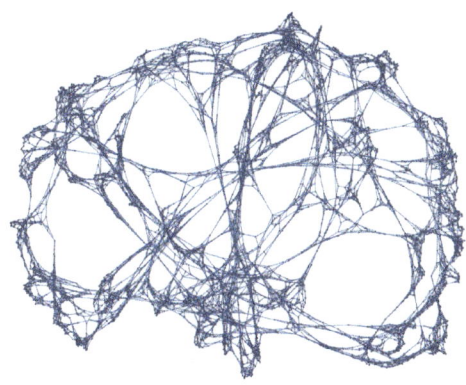

Figure 8: Example of later development of one spatial hypergraph.

And then after a few steps, we're building up this kind of complicated thing (Fig. 8). Now remember this picture that I'm showing is just showing the particular embedding in two dimensions that the Wolfram Language chose in displaying this graph. What is fundamental about this is only the relations, the connections between these atoms of space. [At this point, the speaker presents many visual examples we are unable to reproduce in full here. Fig. 8 is representative. A link to a video of the talk is provided in the preamble.]

Now, one thing we can do is, we can say, well, what's the effective dimension of the thing that we have, i.e., the limit of this big graph?

The way we can estimate dimension is, we say—start at a point, and then just make a ball, effectively, by going to things that are one unit away on the graph, two units, three units, etc., and simply ask the question: How many distinct nodes of the graph do we reach after we do our steps? (See Fig. 9) And if the answer can be approximated by r^d, we say d is the effective dimension of the object-limiting space that emerges from this graph. Things get a bit more complicated. Just like if you were drawing a circle on a sphere, the area of the circle on the sphere is not exactly πr^2. It has a correction term that depends on the overall curvature of the sphere. And so it is with these hypergraphs, you can compute their correction terms, and so on. And you can ask the question: When you look at the limit of one of these things; when you've run it for a long time, what is the overall structure of this? And is there a continuum limit that you can describe for these types of things? Well, it turns out that there is, and it's a little bit like if

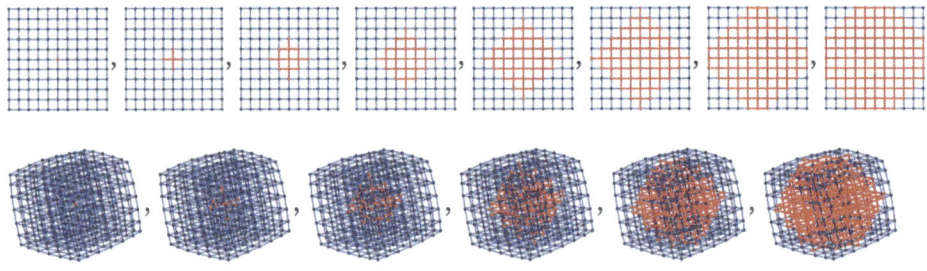

Figure 9: Estimation of the dimension of a spatial hypergraph.

you imagine a bunch of molecules bouncing around and they are colliding and they conserve momentum, and so on, the limit of a bunch of molecules bouncing around the continuum limit is the equations of fluid dynamics.

So what's the continuum limit of this update rule for these atoms of space and this hypergraph? It turns out the continuum limit is Einstein's equations for the structure of spacetime. And so what you can do here is you can derive—there's many more pieces to this whole story—basically, the analogue of the Navier-Stokes equations for fluid dynamics for an updating hypergraph that has this sort of connection information, i.e. the Einstein equations with certain conditions. So a thing to realize is—when we think about: What is the universe made of then?—the first thing to say is: The universe is just this big hypergraph. So in a sense, the universe is just made of space. There's nothing in the universe except space. But there are aspects of space that we interpret as things like electrons and quarks and things like that. It's very much like if you have a fluid like water: There's all these molecules bouncing around. And there may be structures like vortices in the fluid that may have a definite identity and persistence we can identify as definite kinds of objects. And it's the analogue of those kinds of vortices and things that we think of as being things like electrons and so on in this evolving hypergraph. And we can kind of see these analogies, probably between black holes, and things like electrons, in the way that those appear in the structure of this evolving hypergraph. But a little bit disappointingly, it seems like the fraction of the activity of the universe that is all of the things we care about, like all those electrons and photons, things like that, is some tiny 10^{-120} [proportion] or less of all the activity in the universe, the vast majority of the activity of the universe, is devoted to the knitting together of the structure of space, and to maintaining this kind of connection of this hypergraph.

So one of the things you might wonder about in this whole description is: I've described space as the extent of this hypergraph; time is something very different from space in this kind of picture. It is the inexorable computational process of rewriting this hypergraph. So the progress of time is a progress of computation. And one of the things that you can then ask about is, how does relativity work? And how does the relationship between space and time and relativity work? Basically, the answer is what really matters to an observer embedded within this hypergraph.

It's very important that, if we believe this is a model for our whole universe, we are embedded within that model. To an observer embedded within that model, the only thing you can tell is the causal relationships between the update events that happen in this hypergraph. And as soon as you start drawing these causal graphs—and there's a property we call causal invariance—once you have these causal graphs, you will discover that it's inevitable that relativity works for an observer embedded within this hypergraph.

So the next big idea here is: How does something like quantum mechanics work in these kinds of models? Well, it turns out that quantum mechanics is not something where you have to say: Oh, let's add quantum mechanics to the model. Quantum mechanics is an inevitable feature of this model. And I'm going to talk in a moment about how this actually relates quite directly to *Laws of Form* and so on. But let's just talk about quantum mechanics for a second. The big distinction between classical physics and quantum mechanics is in classical physics: we imagine that definite things happen. Throw a ball; it goes in a definite trajectory.

And quantum mechanics? Sort of. The big idea is: No, there isn't a definite trajectory. There are many possible trajectories, and we only get to see a sampling of probabilities across those trajectories. Well, in the way that these hypergraph rewrite systems work, what we define is: there is this rule that says: If you have a little piece of hypergraph that looks like this, you should rewrite it like that. But the point is that we are not saying where exactly to apply that rule. And there are many different places where that rule could be applied. In fact, there are many sequences of different rule applications that could be made. And what we can do is we can make what we call a multi-way graph that represents all possible histories of this hypergraph, corresponding to all possible ways that this rewriting can be done. And so what happens is, sometimes you'll have a particular state of this hypergraph. And the rewriting can be done in two or three different ways or whatever. And you'll have a branch where you can end up with what are, effectively, different states of the universe. Sometimes, two different states of the universe will end up being updated and end up being the same state. So you end up with this kind of branching, merging structure that represents these different states of the universe.

There are much simpler setups. You can just do this with string rewriting systems, for example. You can make a multi-way graph with just string rewriting, where we just have rules for, for replacing A's and B's and so on. We get multiple possible results here, and then they merge and so on (Fig. 10).

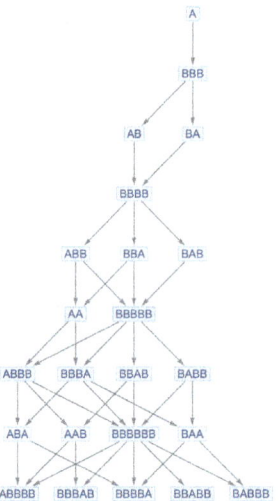

Figure 10: String rewriting system.

What we imagine in our model for physics is that this hypergraph can be rewritten in all these different ways, and we end up with this multiway graph of possible paths of history for the universe (See middle panel, Fig. 6). Well then, the next question is: What do we actually observe in the universe? We've got all these different possible paths of history. What do we actually see happen?

Let's say we slice this (Fig. 6, middle panel, top) to correspond to the possible states of the universe at a particular time. We can make what we call a branchial graph that represents the relationships between those states of the universe. We can simply join two states of the universe if they had a common ancestor on the step before. So in quantum mechanics, this is essentially a map of quantum entanglements. This is a way to represent a quantum state. The choice of what constitutes a particular moment in time has a kind of reference frame issue, just like it does in relativity. If you pick what you consider a particular moment in time, you can produce this branchial graph and you can think of it as a kind of branchial space in which the coordinates in the branchial space represent different possible histories for the universe.

So how come we think definite things happen? Well, it's interesting to see how this works, because in the case of space, we don't think there are little atoms of space. It's new news that there are little atoms of space. Our experience is: Space seems like a continuum. The reason we have that experience is because we're really big compared to the atoms of space, and we are aggregating our experience as observers of this spatial hypergraph. We're looking only at these very aggregated features of it, in particular, down underneath, there's all this computational irreducibility, all of this very complicated behaviour in the detailed atoms of space. But we, as computationally bounded observers of that, all we do is to look at these aggregate properties. And that's why we conclude that space seems like a continuum and so on.

[At this point in the talk, time was getting short, and the speaker presented a menu of possible topics to explore. These can be seen in the video above. We pick up towards the end of the talk.]

So [let us say] you have this rule, and from this very simple rule, you can generate all of this structure that reproduces all sorts of laws of physics. This seems to be a very promising way to understand that physics really is computational all the way down, and it connects with all sorts of other kinds of approaches to mathematical physics. We can enumerate a bunch of those. But one question is, if you think you have the rule and it gives you our physical universe, you're then in this very strange position: Why did we get this rule and not another rule? That seems very mysterious at first. But then you realize that actually, just as we said you were applying this particular rule in all possible ways, we can [further] imagine applying all possible rules in all possible ways. You'd say, what a huge mess that will make—all possible computations done in all possible ways. What you realize is just like what happens with a particular rule, you have something where there's branching and there's merging.

You have this kind of giant entangled object, which is the entangled limit of all possible computations. So, for example, you could take all possible Turing machines started off with all possible initial states. And as they evolve, many of the states they reach will be the same. So you will have this complicated entangled object. We call that object the Ruliad. It's kind of the ultimate limit of sort of computational processes. And what seems to be the case is that we can think of our perception of the physical universe as a particular sampling of this Ruliad object. And so the Ruliad is this entangled limit of all possible computations. But the question is, what can we conclude from that? Well, it turns out for observers like us, it is inevitable that we will perceive certain aspects of that Ruliad. So just as if we have a bunch of molecules bouncing around and we're observers who are large compared to those molecules, and we're observers who don't disentangle all the details of how those molecules are bouncing around, we will conclude that the gas laws are valid.

So, similarly, observers with certain characteristics embedded within this Ruliad will conclude that certain aspects, certain things, are true about about what happens in the Ruliad. So the conditions that we seem to need for observers like us are that we are computationally bounded, and, second, that we believe we are persistent in time. So, you know, at every moment we might be made from different atoms of space, but we believe that we are persistent in time. Those two conditions turn out to be sufficient to give us general relativity and quantum mechanics and not just give us those things approximately. They give us the exact mathematical formalism of general relativity and quantum mechanics. So in a sense, if we were alien observers not like us, we might conclude that the universe works very differently. It's a very different slice of the Ruliad. But for observers like us, we will conclude that general relativity and quantum mechanics work.

So one feature of the Ruliad that I thought I would mention here takes us again back to *Laws of Form*. If the foundation of physics is this Ruliad, which we are sampling as observers, then what about mathematics? Well, we also imagine that the Ruliad is the foundation of all mathematics, the mathematics that we do, and the 3 or 4 million theorems that human mathematicians have written down. Those come from a certain sampling of this Ruliad. The sampling done by a mathematical observer is a bit different from the sampling done by a physical observer. A physical observer is much more oriented towards evolution in time. A mathematical observer is much more oriented towards extension and mathematical space. But it's the same idea, and we can start looking at a kind of meta mathematical space.

Let's imagine we have some relation here:

$$x \circ y \leftrightarrow (y \circ x) \circ y \qquad (2)$$

We can start looking at a multi-way graph that represents the consequences of that relation (Fig. 11).

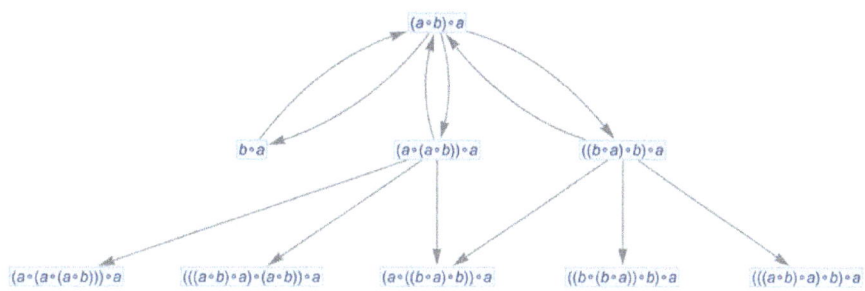

Figure 11: Elaboration of the relation of Equation 2.

We can start thinking about theorems here, where we have the theorem that [the starting orange node on the right] is equal to [the terminal orange node] (Fig. 12). This relation has a proof that corresponds to that path. We can unroll that and say that's the proof. We can do that for many kinds of mathematical structures.

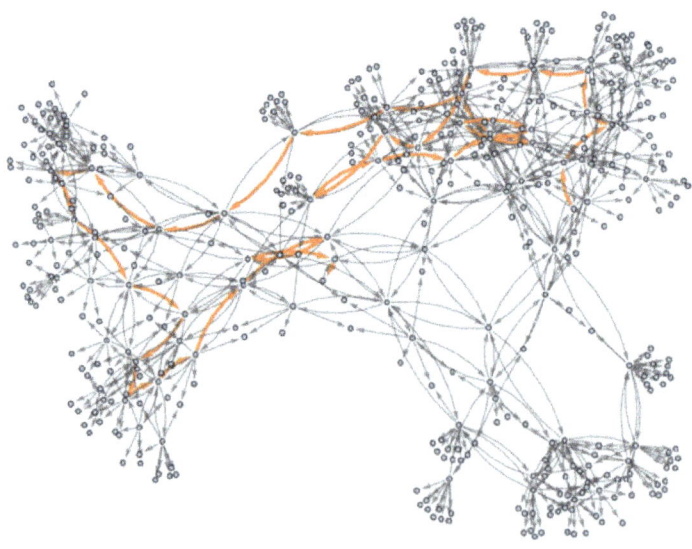

Figure 12: Tracing a proof in a hypergraph of mathematical objects.

We can construct kind of the analogue of bronchial space in Metamathematics, which is this kind of meta mathematical space. It represents the relation between different theorems and mathematics. It's kind of like the Big Bang for metamathematics. We start from this initial condition, and then we derive all these other theorems of mathematics.

[The talk concluded with many visual examples illustrating the notion of the Ruliad and its capacity to contain within it the elements we recognize as mathematical. See the video for details.]

THE MYTH OF THE BOUNDARY—*LAWS OF FORM*, LOGIC AND SETS

LOUIS H KAUFFMAN
University of Illinois at Chicago
`loukau@gmail.com`

Abstract

This article examines the representation of the idea of a distinction via the drawing of a boundary in the contexts of the elements of *Laws of Form,* logic, and a construction of set theory using expressions in the Calculus of Indications restricted to a strong law of calling that is described in Section 3. These explorations extend the reach and understanding of Spencer-Brown's *Laws of Form*.

1 Introduction

In considering the idea of distinction, one can begin by saying that a distinction is drawn by arranging a boundary with separate sides so that a point on one side cannot reach the other side without crossing the boundary. Not every distinction has a visible boundary, but, in fact, the use of such a boundary is one way to represent a distinction. I wish to examine how boundaries are used in these representations and how they lead to semiotic and mathematical relationships. We begin by drawing a circle.

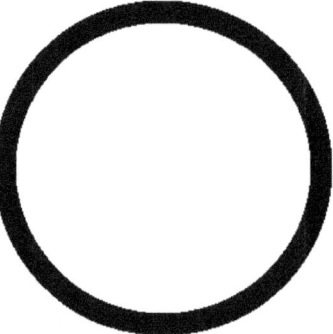

Figure 1: Circle

Consider the circle as "making" the distinction between its inside and its outside in the plane. Notice that this boundary, the circle, is neither on the inside of that distinction, nor is it on the outside. Thus a further distinction has occurred that delineates the boundary circle itself as a third state (neither inside nor outside) associated with the distinction. What role then, does this circle play?

In *Laws of Form* (Spencer-Brown (1969)), one uses a "mark of distinction," an iconic sign for the making of a distinction. Thus, when we write ⌐, this sign can stand for any distinction, and it can stand for the distinction made by the ⌐ in the plane. For the sake of typography, let us use a box instead of the Spencer-Brown mark.

□

This box is taken as an ideograph that is subject to laws of replacement as in Spencer-Brown's laws of calling and crossing.

Calling: □ □ = □

Crossing: ⊡ = .

Recall how these Laws come about.

In calling, each box calls or names the distinction. In fact, each box can regarded as a redundant name for the other box! We do not need two boxes to indicate the one distinction that is indicated by a single box.

In crossing, we regard the box as an instruction to cross from the state indicated on its inside. Thus $\boxed{}$ is an instruction to cross from the marked state that is indicated by a single box. Crossing from the marked state yields the unmarked state.

In this way of thinking, the single box $\square$ makes a distinction in the plane, with the inside unmarked and the outside marked. We take the convention that the mark itself indicates the marked state.

It is remarkable how many interlocking thoughts we have about the distinction made by one box $\square$. G. Spencer-Brown, in *Laws of Form,* uses the mark as a sign in a mathematical language that begins with the Calculus of Indications and the laws of calling and crossing. The representation space is the plane so that one can always refer to the bounded and unbounded sides of the distinction made by any given mark, and so that conventions of written signs can be used. In its inception Laws of Form does not require or use the linear ordering of written text, but later on when writing algebra, it is customary to write the algebra in a linear form, but that is still not necessary for working with LoF. Eventually, after developing the algebra and proving completeness for the algebraic initials, LoF considers, in Chapter 11, recursion and reentry. It is at this point that we return to the boundary in a different way. Consider the equation

$$J = \boxed{J}.$$

We cannot solve this equation in the Calculus of Indications with values marked and unmarked. For if $J = \square$, then the equation defining J says that $\square = \boxed{\square}$, and this is not the case. And if J is unmarked, then the equation defining J says that the unmarked state is equal to the marked state, and this is not so. So far this means that we can write equations in the Primary Algebra of LoF that have no solutions. This should not be surprising since there are equations in ordinary algebra that have no solutions in the real numbers such as x = -1/x. The equation x = -1/x is usually written as $x^2 = -1$ and has a long history culminating in the creation of the complex numbers where every polynomial equation with real coefficients has a solution.

Define {x} = -1/x. Then the equation x = {x} for real numbers is analogous to the LoF equation $J = \boxed{J}$. This leads to the idea that $J = \boxed{J}$ could also have a solution in a new domain beyond the Calculus of Indications.

Let's consider how this idea plays out in Chapter 11 of LoF. There, the equation $J = \boxed{J}$ is interpreted temporally, so that if we rewrote it as $J' = \boxed{J}$, it can be interpreted as saying that the value of J at the next time is $J' = \boxed{J}$. With this temporal interpretation, J is oscillating in time between marked and unmarked values. In the notation of Chapter 11, we can write a reentry circuit as in Figure 2. Then the mark emitted in the outer space of the circuit travels through the wormhole and makes the marker emit an unmarked state the next time. This makes it emit a marked state and so it goes. The equation $J = \boxed{J}$ has the structure of a buzzer – what turns it on, turns it off, and what turns it off turns it on.

The reentering mark J is not directly discussed in Chapter 11, but it is a subtext of that chapter. For example, consider Figure 1 in Chapter 11, which we reinterpret below in our Figure 2.

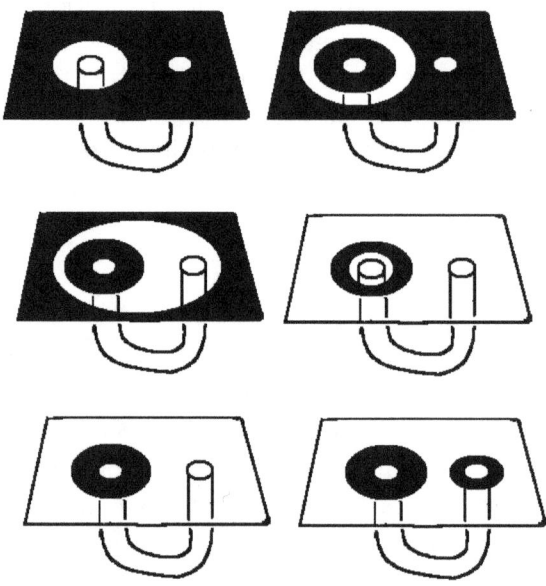

Figure 2: A facsimile of Figure 1 of Chapter 11, *Laws of Form*.

Figure 2 illustrates the temporal action of the reentering mark with a marked state on the outside tunneling to the inside and causing an unmarked state to occur on the outside that feeds back to the inside, *ad infinitum*.

Once again, the boundary is essential as that singular place where the marked state is transformed to an unmarked state and *vice versa*. Once again, we can observe that the boundary is neither marked nor unmarked.

The primary algebraic structure of *Laws of Form* is preserved in the temporal domain. Consider the expression $J\,\boxed{J}$. At any given time, either J is marked or J is unmarked. If the two J's in this expression refer to the value of J at one time, then we see that it is the case that $J\,\boxed{J} = \boxed{}$ no matter what time it is. Now we see that to maintain this form it would not be correct to write $J\,\boxed{J} = JJ$ since this would abrogate the relationship indicated in $J\,\boxed{J}$. This means that we need a new rule of substitution for the use of the equation $J = \boxed{J}$. The rule is what we have previously (Kauffman and Flagg 2019) called the Flagg Resolution, a key insight of James Flagg. The Flagg Resolution proclaims that: *In a given equation, the substitution of $\boxed{J}$ for J must either be not performed, or it must be performed for every occurrence of J in the equation.*

Thus we can write $J\,\boxed{J} = \boxed{J}\,\boxed{\boxed{J}}$, and given that the algebraic version of the law of crossing, $\boxed{\boxed{A}} = A$ (reflection), is allowed we can further write $\boxed{J}\,\boxed{\boxed{J}} = \boxed{J}\,J$. No contradiction occurs, and we see that with the Flagg Resolution, we can continue to use the Primary Algebra with the reentering mark J included. Other reentrant forms can similarly be included in the basic structure of the form and handled according to their temporal interpretations.

Along with the temporal point of view, there is a spatial/notational point of view. Consider an infinite nest of marks:

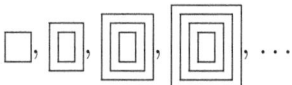

Then it is literally (in the face of an actual infinity of marks) the case that $N = \boxed{N}$. In the temporal view, we have a sequence of nested marks leading up to N as a limiting form.

$$\boxed{},\ \boxed{\boxed{}},\ \boxed{\boxed{\boxed{}}},\ \boxed{\boxed{\boxed{\boxed{}}}},\ \ldots$$

Both points of view are useful, and both are related to the boundary of the first distinction. In the case of the infinite nest N, what is indicated is an infinite oscillation back and forth across the boundary. In the case of the limiting forms, we have finite oscillations of arbitrary length.

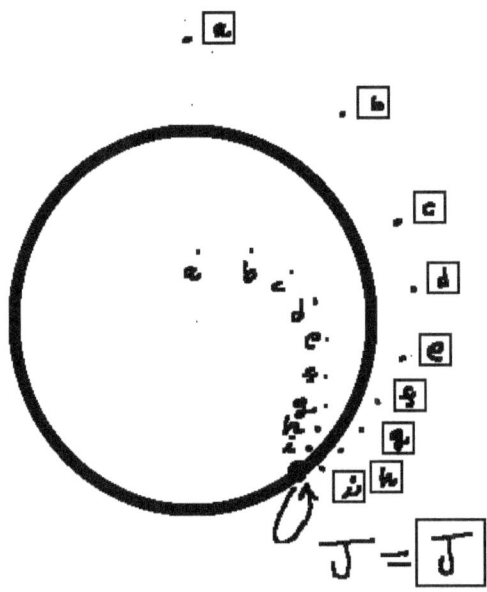

Figure 3: Imaginary boundary Where $X = \boxed{X}$.

We can think of the boundary itself as the reentering mark. Then, points within the boundary are sent by the operation of crossing to the outside, and points outside are sent inside. Points on the boundary do not move under the operation of crossing. They satisfy the equation $X = \boxed{X}$. See Figure 3.

Finally, we return to the mark itself.

$$\square$$

The mark stands in the space in which it makes a distinction between inside and outside. The outside is marked by the mark. The inside is unmarked. The mark itself represents the boundary. The mark is imaginary. The mark is a sign. The role and the genesis of the sign are intimately related to the idea of a boundary.

2 Calculus of Indications, Primary Algebra, Logic and Diagrams.

It is the purpose of this section to recall the Calculus of Indications and the Primary Algebra of LoF. Expressions in the Calculus of Indications are arrangements of marks such that of any two marks in the expression either one is outside the other or neither is outside the other. Thus

is an expression. Any expression can be uniquely reduced to either the marked state or the unmarked state by using repeatedly the laws of calling and crossing. For example, in the expression above we have a reduction to the marked state.

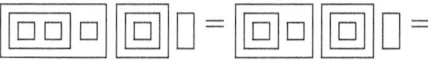

If A and B are two expressions, then we can form a new expression AB by juxtaposing them. If A is a single expression, we can form a new expression by placing A inside a mark to form $\boxed{A}$. These two operations, juxtaposition and crossing, can be used recursively to form all possible expressions.

By letting a symbol, such as A, denote an expression, we have arrived naturally at the level of algebra in relation to the arithmetic of the Calculus of Indications. Since any expression can be reduced to the marked or the unmarked state, we can regard the sign A as indicating the presence or absence of a mark. Algebraic identities are formulas that are true for all possible values of the signs within them. Thus, we have the following:

$$AA = A$$
$$\boxed{\boxed{A}} = A$$
$$\boxed{}A = \boxed{}$$
$$\boxed{A\boxed{A}} =$$
$$\boxed{\boxed{A}\boxed{B}}C = \boxed{\boxed{AC}\boxed{BC}}$$
$$\boxed{A}B = \boxed{AB}B$$

The reader can see by letting A, B and C be either marked or unmarked that these identities hold for every such choice. Note that since $\boxed{\boxed{X}} = X$ for any X, it follows from $\boxed{A\boxed{A}} =$ that $A\boxed{A} = \boxed{}$ for any A.

There are infinitely many such algebraic identities, but one can axiomatize the algebra so that all possible identities of this kind can be derived from one or two initial identities. In LoF, the initials

$$\text{Position: } \boxed{A\boxed{A}} =$$

and

$$\text{Transposition: } \boxed{\boxed{A}\boxed{B}}C = \boxed{\boxed{AC}\boxed{BC}}$$

are shown to be an axiomatic basis for all identities. This is the completeness theorem for the Primary Algebra.

This arithmetic and algebra for distinction can be seen as a basis for logic. For that, one uses the following interpretations:

$$\begin{aligned} \sim A &= \boxed{A} \\ A \vee B &= AB \\ A \wedge B &= \boxed{\boxed{A}\boxed{B}} \\ A \to B &= \boxed{A}B \end{aligned}$$

Here $\sim$ denotes "not," $\vee$ denotes "or," $\wedge$ denotes "and," $\to$ denotes "implies." For truth evaluation, one can take the marked state for True (T) and the unmarked state for False (F). With this choice, note that a juxtaposition AB is T (marked) if and only if either A is marked or B is marked. Similarly, the algebraic expression $\boxed{\boxed{A}\,\boxed{B}}$ is marked (T) if and only if both A and B are marked. The choice for articulating implication is the same as using $(\sim A) \vee B =\sim (A \wedge (\sim B))$. That is, we take "$A$ implies B" to mean that "It is not the case that A is true and not-B is true." We say that "A entails B" and we write $A \to B$.

The structure of implication gives a first opportunity to see how this translation works. Note that

$$\sim (A \wedge (\sim B)) = \boxed{A \wedge \boxed{B}} = \boxed{\boxed{A}\,\boxed{B}} = \boxed{\boxed{A\,B}} = \boxed{A}\,B$$

Many logical identities are transparent in the context of the Primary Algebra. For example we have the identity $A \to B = \sim B \to \sim A$. Writing in Primary Algebra we have

$$(\sim B \to \sim A) = \boxed{\boxed{B}\,\boxed{\boxed{A}}} = B\,\boxed{A} = \boxed{A}\,B = (A \to B).$$

Note that XY = YX is taken as given in LoF since there is no left-right distinction in the plane where the expressions reside.

Here is one more example.

$$(A \to B) \vee (B \to A) = \boxed{A}\,B\,\boxed{B}\,A = \boxed{A}\,A\,\boxed{B}\,B = \boxed{}\,\boxed{} = \boxed{} = T.$$

The structure of this tautology is laid bare. It is valuable to look carefully at such verifications, since there are uses of the term "implies" that would deny the truth of the tautology. The logical form of implication is true exactly when either $\sim A$ is true or B is true. There is no assumption of a causal connection between A and B. A implies B can be false only if it can happen that A is true and B is false. Under this circumstance B implies that A is true.

Logic arises from the structure of a single distinction in conjunction with a multiplicity of variables that can indicate one or the other side of that distinction. When we work with negation, we consider the operation of crossing the boundary of the distinction, from unmarked to marked or from marked to unmarked. When we speak of "A or B," we consider that one or the other of A, B can indicate a marked state. When we speak of "A and B," we consider the possibility that both A and

B can indicate the marked state. When we speak of the truth of "A implies B" we mean that A marked does not co-occur with B unmarked.

It is important to understand that logical operations occur at the relationship of one distinction and the possibility of a multiplicity of distinctions. Two variables afford 4 possibilities. Three variables afford 8 possibilities. A positive integer number of n variables affords a "logical space" of 2^n possibilities.

The Venn diagram (Venn 1881) can be taken as an indication of the possibilities in logical space. The diagrams were invented by Venn to illustrate and facilitate the solution to logic problems, and they accomplish this by producing a diagrammatic logical space that presents all of the different ways that two or more logical variables can be the case with respect to one another.

Consider the two-circle Venn diagram illustrated in Figure 4. In the Venn diagram, each circle makes a distinction between inside and outside. These distinctions can overlap so that A and B share a portion of their insides and indeed all possibilities of sharing are available in the diagram with the convention that $\sim A$ denotes the outside of A and $\sim B$ denotes the outside of B. Then we have the four compartments of the two circle diagram as listed below and illustrated in Figure 4.

$$\begin{aligned}
&\text{The intersection of } A \text{ and } B: & A \wedge B \\
&\text{The intersection of } \sim\!A \text{ and } B: & \sim\!A \wedge B \\
&\text{The intersection of } A \text{ and } \sim\!B: & A \wedge \sim\!B \\
&\text{The intersection of } \sim\!A \text{ and } \sim\!B: & \sim\!A \wedge \sim\!B
\end{aligned}$$

In Figure 5 we have a two-circle Venn diagram labeled using the Primary Algebra of LoF. Thus the intersection of A and B is labeled with $\boxed{\boxed{A}\,\boxed{B}}$. The part of the diagram that is A but not B is labeled with $\boxed{\boxed{A}\,B}$. The part of the diagram that is B but not A is labeled with $\boxed{A\,\boxed{B}}$. The part of the diagram that is neither A nor B is labeled with $\boxed{AB}$.

Working with Venn diagrams gives an extra dimension for understanding Spencer-Brown's algebra. For example, the identity "Extension" in LoF is given by the formula

$$\text{Extension:} \quad \boxed{\boxed{A}\,\boxed{B}}\,\boxed{\boxed{A}\,B} = A.$$

Examine the interior of the circle labeled A in Figure 5, and you will see that it is the

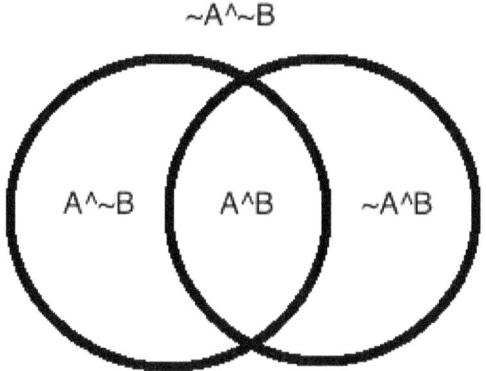

Figure 4: A two circle Venn diagram.

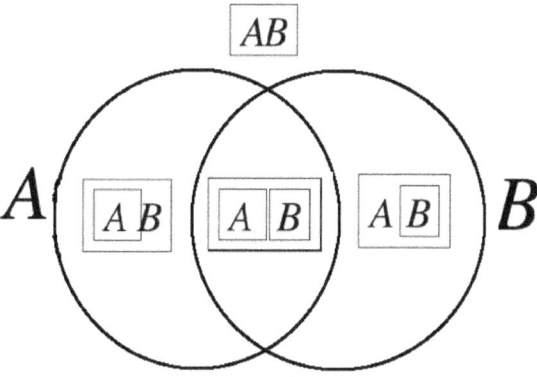

Figure 5: A two-circle Venn diagram labelled in Primary Algebra.

union of the two compartments $\boxed{\boxed{A}\,B}$ and $\boxed{A\,B}$. Thus we see that extension expresses the decomposition of A in the logical space of the Venn diagram.

Similarly, we see that $AB = \boxed{\boxed{A}\,B}\,\boxed{A\,B}\,\boxed{A\,\boxed{B}}$ by noting that the union of all the compartments of A and B together form the right-hand side of this identity. This union identity for A and B is not one of the explicit identities in the book *Laws of Form*, but you can verify it by letting B be either marked or unmarked and seeing that it is so in each of these cases.

The principle behind the Venn diagram takes us back to basics. *Each Venn circle makes a distinction in every other Venn circle by dividing it into two parts. Each Venn circle is a distinction operator that acts on all the other Venn circles in the diagram.* An n-loop Venn diagram shows the ways in which n loops can interact with one another. By making each circle divide the others, the Venn diagram reaches back to the first distinction of a single circle on the one hand while at the same time reaching outward to a multiplicity of distinctions.

The three-circle Venn diagram is the last Venn diagram that can be accomplished with strictly geometric circles. View Figure 7 for a 4-loop Venn diagram. The fourth distinction is represented by a loop that indeed does make a distinction in each of the previous circles, but the fourth loop is only topologically a circle. It is a loop in the plane, making a distinction in the plane. It is a nice puzzle to consider the designs for n loop Venn diagrams. This is an entry point from form and logic to geometry and topology. Figure 8 illustrates a 6-loop Venn diagram. Note how each successive loop has been drawn so that it cuts each previous region into two parts.

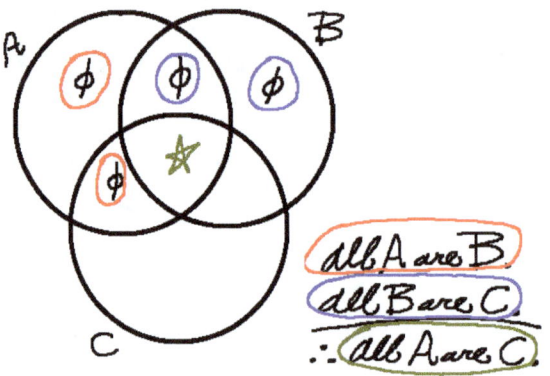

Figure 6: Three-circle Venn syllogism.

The Myth of the Boundary

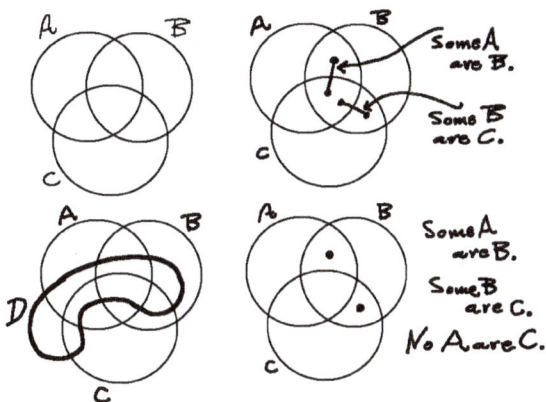

Figure 7: Three-circle Venn diagram.

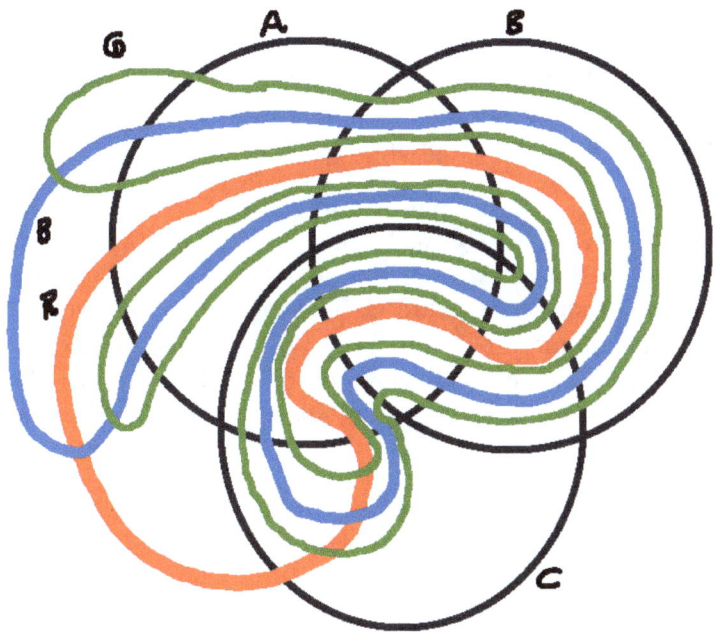

Figure 8: A six-loop Venn diagram.

Consider a syllogism in the standard form

>All A are B.
>All B are C.
>Hence, all A are C.

In the 3-loop Venn diagram of Figure 6, each of the statements in the syllogism has a diagrammatic interpretation. "All A are B" means that any of the A compartments that are not also B compartments must be empty. In the Figure, we have marked empty boxes with a zero. Then Figure 6 shows how if we mark a 3-loop Venn diagram according to the two premises of the syllogism, then, in the resulting marked diagram, we have "All A are C." The Venn diagram is a reliable tabulation of all possibilities that can arise from a set of premises and of how they lead to the conclusion.

The diagrams also allow us to extend our logical reach by including the word "some." We indicate "Some A are B" in the 2-loop Venn diagram of Figure 7 by placing a marker in the form of a bar in the region of intersection between A and B. In a 3-loop Venn diagram, we will indicate "Some A are B" by drawing a segment through the boundary (made by C) between the two regions comprising the intersection of A and B. The segment indicates that at least one of these regions is occupied. With the help of that we can see that the premises

>Some A are B
>Some B are C

do not let us conclude that Some A are C. This is because, as shown in Figure 7, we can have elements common to A and B and other elements in common with B and C, but none of these elements are common to A, B, and C. By illustrating all possible states of affairs, the Venn diagrams allow us to perceive spaces of logical possibility directly.

Figure 9 illustrates the syllogism:

>Some A are B.
>All B are C.
>Therefore, some A are C.

Again, we have used a bar to indicate that one or both of two regions may be occupied, but then the stricture that all B are C tells us that one end of the bar

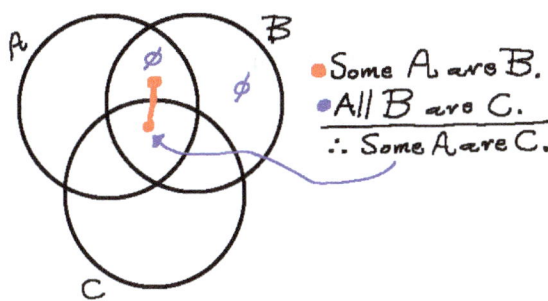

Figure 9: Another syllogism.

is definitely not occupied, and so, the other end must be occupied, and this is the conclusion of the syllogism.

This introduction to syllogisms by way of Venn diagrams is a good way to appreciate how basic logic works, and it is a way to see how the arithmetic, logic, and algebra of a single distinction of LoF apply to a multiplicity of overlapping distinctions.

This Venn approach to syllogism should also be compared with the highly linguistic approach of Spencer-Brown to syllogisms, as it is found in Appendix 2 of *Laws of Form*. In that appendix, Spencer-Brown first writes the form of the standard syllogism BARBARA

$$A \to B$$
$$B \to C,$$
$$\text{Hence, } A \to C.$$

Then, this is translated into Primary Algebra.

In symbolic logic, we have:

$$(A \to B) \wedge (B \to C) \to (A \to C)$$

This, in turn, becomes the formulas below in LoF.

Note that this final form of the statement of the syllogism BARBARA has the form $\boxed{P}\boxed{Q}\,R$ where P and Q are premises P = (A implies B) and Q=(B implies C), and R is the conclusion R=(A implies C).

Spencer-Brown points out that we can obtain other correct syllogisms from BARBARA by permuting its form without changing its truth value. This includes replacing any chosen variable by its negation in all places of its appearance in the formula for the syllogism. His method works at the symbolic level without actually evaluating the expressions. There are 24 such rearrangements of BARBARA. These comprise all valid syllogisms that involve some and all where we work under the assumption that when we say "All A are B," it is possible that A is empty, and we regard "All A are B" as true when A is empty. However, when we say "Some A are B," it is assumed that A is not empty.

Here is a rearrangement of the premise/conclusion form that we have articulated. The initial form is a syllogism with premises P, Q, and conclusion R.

$$\boxed{P}\boxed{Q}\,R = \boxed{P}\boxed{R}\,Q.$$

We conclude from permutation that the syllogism with premises P and ~R will have conclusion ~Q. Using P, Q, and R from BARBARA this becomes

$$\boxed{A\,B}\boxed{A\,C}\,B\,C$$

And this is the syllogism:

    All A are B.
    Not all A are C. = Some A are not C.
    Hence, Not all B are C. = Some B are not C.

    All A are B.
    Some A are not C.
    Hence, Some B are not C.

We leave it as an exercise for the reader to draw a Venn diagram showing the validity of this syllogism!

Remark. Note that we have Some A are B = Not all A are not B = $\boxed{A\,B}$. We also leave it as an exercise for the reader to make translations of other some and all statements into Primary Algebra.

Remark. Consider the invalid syllogism:

Some A are B,
Some B are C,
"Hence" some A are C.

In premise-conclusion form for LoF this is:

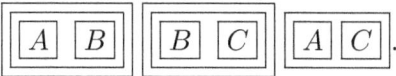.

It is not hard to see that this form cannot be obtained from BARBARA by permutations, replacements or uses of reflection ($\boxed{\boxed{X}} = X$). These are Spencer-Brown's rules for handling the syllogistic form. One needs to be warned that this form, which is not a valid syllogism, does reduce to the marked state. Thus algebraic handling of these forms is highly restricted. Appendix 2 of *Laws of Form* has a discussion about why this is so, and the reader can see the author's discussions of this point in (Kauffman 2013). In this paper, I suggest the point of view that Spencer-Brown's structural approach should be regarded in the larger context of multiple distinctions and the calculus of Venn diagrams.

The Spencer-Brown permutation calculus (described above) intermediates between the Calculus of Indications (where the universal and the particular coincide) and the Venn diagrammatic where a multiplicity of distinctions interact. It is remarkable that logic involving the distinction of that which is particular (as in the use of the word 'some') and that which is universal (as in the use of the word 'all') can be articulated at this interface, informing both the condensed and expanded worlds of discourse.

The boundary calculus of the Venn diagrams provides a background of profound clarity in working with the subtle relationships of the universal and the particular. Spencer-Brown's insight shows that the form of the syllogism BARBARA implicitly carries the entire story for the use of the logical words 'some' and 'all.' In this sense we see the relationship between the properties of the single first distinction and the logic of multiplicities that are related to it.

The subject of classical syllogisms has a wider class of forms than just BARBARA and its permutations. See (Łukasiewicz 1957) for an accounting of these classical forms. The multiplicity of classical syllogistic forms arises from assumptions that take given possibilities in logical space as non-empty. For example, one has the classical form

BARBARI.

Given: A is not empty.

All A are B.
All B are C.
Hence, Some A are C.

Since A is not empty, we can say "Some A are C", which assumes the existence of elements of A. With the assumption that A is not empty, we can write a primary algebraic version of BARBARI:

$$\boxed{\text{All } A \text{ are } B.}\;\boxed{\text{All } B \text{ are } C.}\;\text{Some } A \text{ are } C. = \boxed{\boxed{A}\;B}\;\boxed{\boxed{B}\;C}\;\boxed{\boxed{A}\;\boxed{C}}$$

We can then make permutations and substitutions on this form. For example, we can replace B by $\boxed{B}$ and permute the last two boxes, to obtain:

$$\boxed{\boxed{A}\;B}\;\boxed{\boxed{A}\;\boxed{C}}\;\boxed{BC} =$$

$$\boxed{\text{All } A \text{ are not } B}\;\boxed{\text{All } A \text{ are not } C}\;\text{Some not } B \text{ are not } C$$

Thus it is claimed that the syllogism below is valid:

(Given: A is not empty.)

No A are B.
No A are C.
Hence, there are objects that are neither B nor C.

This tells you that the only possible elements of A are neither in B nor in C from the premises. Since A is not empty there must be some A that are neither B nor C.

The LoF method spins out 24 syllogisms related to BARBARI. One can do a similar analysis of all the classical syllogisms. A more complete description of this classification will appear elsewhere.

3 LoF Sets

As we move into the possibility of multiple distinctions with Venn diagrams, relationships with collections are natural.

The beginning of set theory (See Kanamori 2007) arises naturally from LoF by *banning the law of crossing and keeping the law of calling in the strong form that* $AA = A$ *for any expression* A.

With this stricture we have, for example,

$$\boxed{\square\,\boxed{\square}\,\square} = \boxed{\square\,\boxed{\square}\,\square} = \boxed{\square\,\square}$$

and the right-most form is reduced since two nested marks do not cancel. No further applications of strong calling are possible. There are infinitely many distinct forms in this "calling calculus." For example, here is an infinite list of distinct forms when we can only use strong calling.

$$\square,\ \boxed{\square},\ \boxed{\boxed{\square}},\ \ldots$$

Definition. *Call an expression whose shallowest depth is a single box a LoF set.* Thus, the above expressions are LoF sets.

Note that any LoF expression is a juxtaposition of such "boxes." We call the members of a LoF set, those LoF sets that are juxtaposed on the inside of its outer box. Thus the members of $\boxed{\square\,\boxed{\square}\,\square}$ are $\square$ and $\boxed{\square\,\square}$. The first member has no members. The second member has, itself, two members.

Note also that if E is any expression in LoF, then $\boxed{E}$ is a (possibly unreduced) LoF set. For example if $E = \boxed{\square\,\boxed{\square}\,\square}\,\boxed{\square\,\square}$, then $\boxed{E} = \boxed{\boxed{\square\,\boxed{\square}\,\square}\,\boxed{\square\,\square}} = \boxed{\square\,\boxed{\square}\,\square}$ forming a LoF set with three members. In ordinary set theory notation we write $\{a, b, c\}$ where $a, b,$ and c are sets. Then, if $a, b,$ and c are LoF sets, then the corresponding LoF set is $\boxed{abc}$, and we may write

$$\{a,b,c\} = \boxed{abc}$$

In representing LoF sets, we do not need to write any parentheses, since the boundaries of the forms delineate the members.

$\boxed{\boxed{}\boxed{}}$ is a LoF set, and its members are the LoF sets $\boxed{}$ and $\boxed{}$.

Note that $\boxed{}$ is the empty set and corresponds to the usual notation $\{\}$ for the empty set. $\boxed{\boxed{}}$ is the set whose member is the empty set and corresponds to $\{\{\}\}$. $\boxed{\boxed{}\boxed{}}$ = $\{\{\},\{\{\}\}\}$, where I have used the comma in the standard set notation.

In set theory, two sets are *equal* if they have exactly the same members. Our adoption of the strong law of calling makes it a

Theorem. *Two LoF sets are equal if and only if they have the same members.*

The reader may enjoy verifying this result. We will give a detailed proof of it elsewhere.

An example will help.

Note that $\boxed{\boxed{}\boxed{}} = \boxed{\boxed{}}$, and these indeed have the same member $\boxed{}$. The repetition of $\boxed{}$ in the left set can be reduced by strong calling. Repetition of membership is reduced on the LoF side via the use of calling. If we had started with multi sets such as $\{a,a,a\}$ where many copies of a member are allowed, then we get set theory by identifying identical members. Thus, as sets $\{a,a,a\}=\{a,a\}=\{a\}$. Such identification occurs inside the members, since they also represent sets. Thus for example we have $\{\{a,a,b\}, \{a,b\}, \{c\}\}=\{\{a,b\}, \{a,b\}, \{c\}\}=\{\{a,b\},\{c\}\}$. In LoF sets, this reduction corresponds to applications of strong calling.

$$\boxed{\boxed{aab}\boxed{ab}\,c} = \boxed{\boxed{ab}\boxed{ab}\,c} = \boxed{\boxed{ab}\,c}$$

There is an exact correspondence between reduction via strong calling and the dictum that sets are equal when they have the same members.

LoF sets comprise a model for the finite sets that can be constructed by repeated collection operations from the empty set. This is actually quite a useful fragment of set theory.

The Myth of the Boundary

We will develop LoF sets in more detail elsewhere and also extend them to infinite sets via infinite forms and reentry forms in the LoF context. The reader may enjoy experimenting with infinite sets via infinite forms. Thus

$$N = \{0,1,2,\ldots\} = \{\{\},\{\{\}\},\{\{\},\{\}\},\ldots\} = $$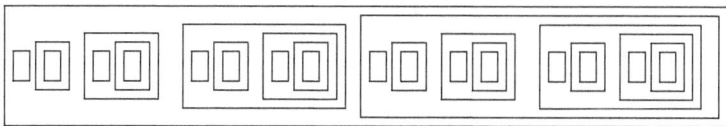

This infinite set represents the von Neumann construction of the natural numbers with

0={}
1={{}}
2={{}, {{}}}
3 = {{}, {{}}, {{}, {{}}}}
and
n+1 = {0,1,2,3,...,n}.

It is interesting to build these sets as forms. Here, for example, is the representative of the number 5.

A new appreciation of the von Neumann construction arises as we see 5 as a LoF set with five members that is in reduced form via strong calling.

Ordered Pairs

We end with a brief discussion of *ordered pairs*. Set theory does not distinguish the order of the elements of a set. Thus $\{a,b\}=\{b,a\}$. An early accomplishment in set theory due to Kuratowski (1921) was to give a set theoretic definition of the concept of an ordered pair (a,b) where $(a,b)=(c,d)$ if and only if $a=b$ and $c=d$. The challenge was to define a set (a,b) depending on two sets a and b so that this order property was a consequence of the construction. Kuratowski's elegant solution was to define (a,b) by the formula $(a,b)=\{\{a\}, \{a,b\}\}$.

It is a bit of fun to see that Kuratowski's definition works. First of all, suppose that a and b are distinct sets and that c and d are distinct sets. Then if $\{\{a\}, \{a,b\}\}=\{\{c\}, \{c,d\}\}$ we conclude that $\{a\}=\{c\}$ and $\{a,b\}=\{c,d\}$

since $\{a\}$ has one member and so cannot be equal to $\{c,d\}$ which has two members. Since $\{a\}=\{c\}$ we conclude that $a=c$ (sets that are equal have the same members). Since $a=c$, it follows that $b=d$ since we know that $\{a,b\}=\{c,d\}$. This checks the property for ordered pairs in this case.

Now suppose $a = b$, then $(a,b)=(a,a)=\{\{a\},\{a,a\}\}=\{\{a\},\{a\}\}=\{\{a\}\}$, so (a,a) has only one member. This tells us that if $(a,b)=(c,d)$ and $b = a$ then $c = d$, since otherwise (c,d) would have two members while (a,a) has only one member. With $a = b$ and $c = d$ we have that $(a,b)=(c,d)$ if and only if $\{\{a\}\}=\{\{c\}\}$ and this happens if and only if $a = c$. This completes the proof that the Kuratowski construction has the ordered pair property.

Does the Kuratowski ordered pair go over for LoF sets? We have to consider the possibility of void entries in the pair. There is no void in set theory, but we have the unmarked state in LoF. We want to have (,), (, □), (□,), (□, □) as distinct ordered pairs. This also means that along with the empty set □ we want to have the *void set*, represented by the unmarked state. The void set is not surrounded by anything, but in an ordered pair, it should become an empty space. We can try the direct transcription of the Kuratowski definition as shown below. As you can see, we will then have that (□,) = (□, □) and we do not want that!

$$(a, b) = \boxed{\boxed{a}\boxed{ab}} \quad (Kuratowski)$$

$$(,) = \boxed{\boxed{}\boxed{}} = \boxed{}$$

$$(\Box,) = \boxed{\boxed{\Box}\boxed{\Box}} = \boxed{\Box}$$

$$(, \Box) = \boxed{\boxed{\Box}\boxed{\Box}}$$

$$(\Box, \Box) = \boxed{\boxed{\Box}\boxed{\Box\Box}} = \boxed{\Box}$$

So here is an answer to putting Kuratowski into the form. We define (a,b) as shown below and as the reader can see. The four basic ordered pairs are now distinct from one another.

The Myth of the Boundary

$$(a,b) = \boxed{\boxed{a}\,\boxed{\boxed{a}\,\boxed{b}}}$$

$$(\,,\,) = \boxed{\boxed{\boxed{}}\,\boxed{\boxed{}\,\boxed{}}} = \boxed{\boxed{}}$$

$$(\boxed{},\,) = \boxed{\boxed{\boxed{\boxed{}}}\,\boxed{\boxed{\boxed{}}\,\boxed{}}}$$

$$(\,,\boxed{}) = \boxed{\boxed{\boxed{}}\,\boxed{\boxed{}\,\boxed{\boxed{}}}}$$

$$(\boxed{},\boxed{}) = \boxed{\boxed{\boxed{\boxed{}}}\,\boxed{\boxed{\boxed{}}\,\boxed{\boxed{}}}} = \boxed{\boxed{\boxed{}}}$$

It may seem like a great deal of work to find this definition of an ordered pair for sets in LoF, but it is satisfying to see that ordering can emerge intrinsically from the unordered world of the calculus of LoF sets.

From this point of view of the world of sets, the notion of membership is seen to be peripheral. The sets are built from successive distinctions using only the strong law of calling. The act of collection is seen as an act of distinguishing an expression in the form. All sets arise from such acts of distinction.

It now becomes possible to re-examine many issues that have been perplexing in the set theory. For example, can a set be a member of itself? In the LoF set context, the simplest set S that could be a member of itself would satisfy the equation $S = \boxed{S}$. We see that such an equation can only be satisfied by an infinite nest of boxes, as in $S = \boxed{\boxed{\boxed{\cdots}}}$. Thus, sets that are members of themselves are outside the constructions based on finite expressions in LoF. If we extend to infinite expressions, then there will be LoF sets that are members of themselves. They need not be banished from consideration.

Propositions and the Russell Paradox.

We are used to defining sets by propositions. For example, let $E(x)$ denote the statement "x is an even number." Then $\{x|E(x)\}$ refers to the set of even numbers, and in LoF sets this is

$$\{0,2,...\} = \boxed{\square\,\boxed{\square}\,\boxed{\boxed{\square}}\,\cdots}\,,$$

using the constructions we have used above. The interesting boundaries of this method of defining sets are well known. For example, we can write $R = \{x | x \text{ is not a member of itself}\}$ (the Russell set). To make R into a LoF set you would need to juxtapose all the finite reduced (by strong calling) expressions with one outer box. None of these are members of themselves. In so doing, you would produce an infinite expression R. Since 'x is not a member of itself' did not specify that x is finite, we can ask if R is a member of itself. But by our construction, R is not a member of itself! So we should include R in the infinite juxtaposition. You can see there is a problem here. First we had $R = \boxed{S}$ where S is the infinite juxtaposition of all the finite LoF sets. Now the definition of R suggests that we should have $R = \boxed{SR}$, but this is a recursive definition! Russell regarded R as a paradoxical set since it appeared that it should both be a member of itself and not be a member of itself 'at the same time.' But if we try to build R we find

$$R = \boxed{SR} = \boxed{S\,\boxed{SR}} = \boxed{S\,\boxed{S\,\boxed{SR}}} = \boxed{S\,\boxed{S\,\boxed{S\,\boxed{S\,\boxed{S\,\boxed{S\,\boxed{S\,\boxed{S\,\boxed{SR}}}}}}}}} = \cdots$$

Each successive stage in the recursion has the indicated $R = \boxed{S}$. So that we have the approximations:

$$\boxed{S\,\boxed{S}} \Rightarrow \boxed{S\,\boxed{S\,\boxed{S}}} \Rightarrow \boxed{S\,\boxed{S\,\boxed{S\,\boxed{S\,\boxed{S}}}}} \Rightarrow \boxed{S\,\boxed{S\,\boxed{S\,\boxed{S\,\boxed{S\,\boxed{S\,\boxed{S\,\boxed{S}}}}}}}} \Rightarrow \cdots$$

None of the approximations is a member of itself. The approximations do tend to the limit

$$L = \boxed{S\,\boxed{S\,\boxed{S\,\boxed{S\,\boxed{S\,\boxed{S\,\boxed{S\,\boxed{S\,\boxed{\ldots}}}}}}}}} \quad \text{with } L = \boxed{SL}.$$

We can regard the Russell set as always approximating a set that would be a member of itself but never succeeding in finite time. This is one way that we can construct a resolution to the Russell Paradox in terms of LoF sets.

With this remark, we will end our introduction to LoF set theory and encourage the reader to continue the exploration on its own grounds.

4 Epilogue

The theme of this essay has been the representation of distinction in terms of boundary. In the introduction and the second section, we discussed the structure of the Calculus of Indications, its Primary Algebra for *Laws of Form* and its relationship with logic. *Laws of Form* is a study of the possibility of a first distinction. Logic, with its multiplicity of variables is concerned with the interaction of many distinctions. *Laws of Form* and logic meet in the realm of Venn diagrams where distinction boundaries cross one another and where two distinctions act to make divisions in each other. We have demonstrated the direct relationship of the Venn diagrams and the Primary Algebra of *Laws of Form*, giving a wider view for the interaction of logic, distinction and boundary. We have extended Spencer-Brown's linguistic analysis of syllogisms to include classical syllogisms with different assumptions of existence. This was illustrated with the syllogistic form BARBARI. In the third section, we showed how a significant fragment of set theory can be constructed from the expressions in the Calculus of indications subjected only to the generalized law of calling: $AA = A$. In this set theory, the empty mark $\square$ represents the empty set. In *Laws of Form* we have the unmarked state as well as the marked state, and so in this set theory, there is also an unmarked state. The empty set has no members. The unmarked state has no boundary within which to search for members. Set theory arises from framing the unmarked state. Indeed, the form we take to exist arises from framing nothing.

It should be remembered that *Laws of Form* is about only one distinction. In its calculus, the universal and the particular coincide. Our explorations into multiplicity should be viewed from this unity and seen in the light of a single distinction between what is marked and what is unmarked or void.

References

[1] Kanamori, A. 2007. *Set Theory From Cantor to Cohen*. Elsevier.

[2] Kauffman, L. H. 2013. Laws of form and topology—Presentation and discussion. *Cybernetics and Human Knowing*, 20(3–4): 50—100.

[3] Kauffman, L. H. and Flagg, J. M. 2019. The Flagg resolution revisited, *Cybernetics and Human Knowing*, 26(2–3): 87–106.

[4] Kuratowski, C. 1921. Sur la notion de l'ordre dans la Théorie des Ensembles. *Fundamenta Mathematicae*. 2(1): 161—171.

[5] Łukasiewicz, J. 1987. *Aristotle's Syllogistic from the Standpoint of Modern Formal Logic*. New York: Garland Publishers, first published 1957.

[6] Spencer-Brown, G. 1969. *Laws of Form*, London: George Allen and Unwin Ltd.

[7] Venn, J. 1881. *Symbolic Logic*, London: McMillan & Co.

Cellular Laws of Form

Kevin German

indicate@kevingerman.de

Abstract

George Spencer-Brown's *Laws of Form* (LoF) provides important insights and inspiration for all sorts of fields. However, his work often takes place on a purely formal level, although the thoughts are inherently carried by space and form. In this work, we present a set of rules for a cellular automaton (CA) that can be used to simulate LoF in an intuitive visual way. By simulating LoF, the functionally complete CA presented here is also able to compute logic, algebra, logic circuits and can be used to build uniform circuits of arbitrary size. It is shown that a CA is a viable and obvious representation of LoF that enables one to simulate the system dynamics and parts of the re-entry simply and visually. Moreover, this demonstrates that LoF can also emerge solely based on a few rules in a many-particle system. The emergence of LoF by a CA could be of philosophical interest. In addition, this CA allows one to gain an intuitive visual understanding of the underlying dynamics.

1 About Laws of Form

With his work *Laws of Form* published in 1969 (Spencer-Brown 1979), George Spencer-Brown provided fundamental thoughts and a viable formalism for second-order observations, cybernetic biology and radical constructivism. His thoughts on the most basic operation of distinction and re-entry are central to Luhmann's systems theory of social systems (Luhmann 2006, 39–43). It also lays a foundation and computational possibility for Varela's and Maturana's notion of Autopoiesis (Varela, 1975; Varela and Gougen, 1978) In addition, (radical) constructivism utilizes Spencer-Brown's insights to explain the constructive nature of one's reality (Watzlawick 1984, 252–255; 330–331). LoF also inspired mathematics (Kauffman 2013).

LoF tries to derive mathematics, logic, and parts of the philosophy of mind (Spencer-Brown 1979, xix–xxix) based on an operation that is both operand and operator. This operation, which can also be understood as a form, is the so-called distinction. Under the basic operation of the distinction, one can understand exactly what we do every day. It is the creation of a border as well as the creation of two sides—the separated and the separated from (here it already becomes linguistically apparent that a clear assignment of the concept of separateness is not possible), the negative and the positive, the inside and the outside. Furthermore, it is worth noting that every indication comes with a distinction and *vice versa*. In order to indicate something as something, it must be distinguished, and in order to distinguish something, it must be indicated (Spencer-Brown 1979, 1–7). Probably the simplest representation of a distinction (also used by Spencer-Brown) is a circle with an infinitesimal border, which is to be equated with its border and at the same time the inside and outside (ibid., 70). We will use, to explain the rules, the compact notation of the distinction introduced by Spencer-Brown, that represents the corner of a rectangle (a form that also distinguishes an inside and outside) (ibid., 4). The CA will use circles.

$$\neg$$

Like every system, this form is also subject to axioms or laws. Since they are axioms, one cannot derive them, but one can still try to get an intuitive understanding. The first axiom can be summarized in Spencer-Brown's words as follows: "[t]he value of a call made again is the value of the call." (ibid., 1) Pointing to something and calling it the same name twice or distinguishing the same thing, based on the same criteria of distinction, does not change the observation. This is the Law of Calling (I1) that can be expressed in our formal notation as (ibid., 5)

$$\neg = \neg\,\neg \tag{I1}$$

The second axiom can be summarized as "[t]he value of a crossing made again is not the value of the crossing." (ibid., 2). To distinguish something as something and to make a further distinction based on the first distinction does not lead to the same distinction. Every indication is at the same time also a border crossing. But to cross a border twice, once from the outside to the inside and then from the inside to the outside, leads again into the initial space (ibid., 2). This so-called Law of Crossing (I2), can be expressed in our formal notation as (ibid., 5).

$$\overline{\overline{}\,}\ =\ \qquad\qquad (I2)$$

From only these two axioms (I1, I2), a variety of theorems and insights emerge. For example, the algebraic consequence (through Theorems 8 & 9 as initials) where a and b stand as placeholders for possible distinctions (ibid., 25).

These distinctions always take place in space. If one were to draw a circle and thus a distinction on a white sheet of paper, the space would be the background on which the distinction is made. The distinction creates a new space that is separated from the original one. We denote a space that contains a distinction, or cross, as marked, otherwise unmarked (ibid., 4–5). It can be shown that any expression of the primary arithmetic can be reduced to one of these two values by applying the laws recursively (ibid., 12–13). The following equivalences show how basic logical terms can be transformed into LoF (ibid., 114). The final truth term is obtained by complete reduction to one of the two states. We denote a marked state as True and an unmarked state as False (ibid., 113).

$$\overline{} \equiv \neg a \qquad\qquad \text{(NOT)}$$

$$ab \equiv a \vee b \qquad\qquad \text{(OR)}$$

$$\overline{\overline{a}\,\overline{b}} \equiv a \wedge b \qquad\qquad \text{(AND)}$$

$$\overline{a}\,\overline{b} \equiv \neg(a \wedge b) \qquad\qquad \text{(NAND)}$$

$$\overline{a}\,b \equiv a \Rightarrow b \qquad\qquad \text{(Implication)}$$

$$\overline{\overline{a}\,\overline{b}}\,\overline{ab} \equiv a = b \qquad\qquad \text{(Equality)}$$

These expressions can be combined arbitrarily, thus we also see that it is no longer possible to distinguish meaningfully between operation, operand and the result. (Remember a & b denote placeholders for potential crosses.) From the above equations, it can be seen that LoF is functionally complete. For this, it is sufficient to see that NAND has an equivalent representative (equation: NAND). It is known that NAND is a Sheffer function that is able to produce any truth table (Wernick,

1942). This means any truth table and logical function can be expressed in LoF. Probably the most staggering idea that emerges from LoF is that of the so-called re-entry. We will cover that later. We will show that all the properties of LoF established above can be replicated using a CA. This will also show that the CA shares most the properties of LoF including functional completeness.

2 About Cellular Automata

Cellular automata are mathematical models that can be used to model complex discrete system dynamics, based on relatively simple rules. A classical 2D CA consists of a space of discrete cells, arranged in a grid (imagine a chessboard). Each cell at position i, j has at a specific time step t a state $a_{i,j}^t$ from an alphabet S and a specific neighborhood $N(a_{i,j}) = \{a_{i,j}, a_{i-1,j}, a_{i+1,j}, a_{i,j-1}, a_{i,j+1}\}$. In this case, N consists of 4 neighbors arranged in a cross and the observed cell in the center itself—also called Von Neumann neighborhood. Another common neighborhood is the Moore neighborhood, which consists of 9 cells, arranged in a 3 by 3 square. Each cell $a_{i,j}^t$ and its neighborhood $N(a_{i,j}^t)$ is transformed into a new state $a_{i,j}^{t+1}$ according to a transfer function ϕ such that $\phi(N(a_{i,j}^t)) = a_{i,j}^{t+1}$ (Wolfram 1985, 901–902). For convenience, let ν represent the neighborhood of a without a, thus $\nu = N(a) \setminus a$.

Invented by John von Neumann (von Neumann 1966; Wolfram 2002, 879), CA are probably best known today through Conway's Game of Life, in which each cell has two states, dead or alive; $S = \{0, 1\}$ and a Moore neighborhood. The next state is determined in accordance with three rules: 1) Any cell with two or three neighbors lives on; 2) A dead cell with three living neighbors becomes alive; 3) All other cells remain dead.

$$\phi(a) = \begin{cases} 1 & \text{if } a = 1 \land 2 \leq (\sum_{n \in \nu} n) \leq 3 \\ 1 & \text{if } a = 0 \land (\sum_{n \in \nu} n) = 3 \\ 0 & \text{else} \end{cases}$$

For the next time step/tick, all rules are applied to all cells simultaneously, and they are updated accordingly as in Fig. 1 (Gardiner 1970). Countless complex dynamic structures emerge from these simple rules. It can even be shown that several of these rules can simulate a Turing machine (Ollinger 2008). CA are nowadays used, e.g. in biology, computer science, or physics (Das 2012, 758–760). Noticeable are the results of CA that generate random sequences with no recognizable pattern from simple rules. For example, Wolfram's Rule 30 (Wolfram 2002, 27–31). CA are thus

1	0	0	0		1	1	0	0		0	1	0	0		0	1	1	0
1	1	0	0	$\xrightarrow{t+1}$	0	0	1	0	$\xrightarrow{t+2}$	0	1	1	0	$\xrightarrow{t+3}$	0	1	0	1
1	1	1	0		0	0	0	1		0	0	1	1		0	1	1	1
1	1	1	1		1	0	0	1		0	0	0	0		0	0	0	0

Figure 1: Conway's Game of Life simulated for 3 steps with a random starting pattern. 1 stands for living and 0 for dead cells.

the prime example of how complexity can emerge only from simple rules and local interactions.

Besides this parallel of generating complexity from simple rules, there is also a deep philosophical connection between LoF and CA. The idea that the universe is at its core a CA is not uncommon today. First introduced in 1969 by Konrad Zuse in his book *Calculating Spaces* (Zuse 1969), the idea was continued and refined by Stephan Wolfram in his 2002 published work *A New Kind of Science* (Wolfram 2002), an attempt to simulate nature-analog processes based on CA. Thus, every cell of the universe with a certain neighborhood is to be transferred into a new state on the basis of transfer rules, our laws of nature (Wolfram 2002, 465–474). George Spencer-Brown's work, which was published in the same year as Konrad Zuse's thoughts on CA, has a similar philosophical implication. The basal operation from which the universe emerges, according to Spencer-Brown, is the distinction, through which "[...] a universe comes into being when a space is severed or taken apart. The skin of a living organism cuts off an outside from an inside. So does the circumference of a circle in a plane" (Spencer-Brown 1979, xxix). Moreover, it should also be clear, almost trivial, that the basal operation that enables us to recognize/experience something at all is the distinction. In order to recognize a letter, objects, persons or ideas, these must be distinguished from the background on which they occur. Basic frameworks regarding discrimination, as a basic cognitive operation, have already been researched (Heylighen 1992). It should be apparent that a CA that produces LoF and the distinction might be philosophically interesting. Highly exciting thoughts on LoF and CA have already been made by Isaacson and Kauffman regarding recursive distinctioning (Isaacson and Kauffman 2016). In the following, other rules are derived which aim at the faithful visual representation of Spencer-Brown's thoughts in a CA.

3 Laws of Laws of Form

In order to derive the rules for the CA, let us list one by one the basic observations on LoF and what kind of rules would be needed to reproduce them. We denote the cells of a cross that encloses an unmarked space (US) ⌐ as unmarked cross cells (UC). Similarly, the cells of a cross that encloses a marked space (MS) ⌐⌐ are marked cross cells (MC). We use the Von Neumann neighborhood because it allows for simpler rules. A Moore neighborhood is still possible but leads to more complex edge cases for the circular crosses.

1. A UC marks the outside space. Expressed as a rule for a CA, this can be stated as follows: If a neighbor is a UC, this cell is marked.

$$\phi_{US}(a) = \begin{cases} MS & \text{if } \exists n(UC(n) \land n \in \nu) \\ US & \text{else} \end{cases}$$

2. It can be seen by I2 that the MC does not mark the space. However, the cells may be marked by another UC in space. Thus it could be seen as an identity element: ⌐⌐ = . and ⌐⌐⌐ = ⌐ . In this sense cells ignore MC as neighbors and no rule is needed.

3. Any space on which a UC appears is marked, regardless of what state it was in before. In this sense, MS displaces US. Preliminary CA rule: If any neighbor of a US is marked, then this cell is also marked. Together with Rule 1, it can be seen that any space can be filled as marked when a UC appears. One can imagine that the markedness spreads "virulently" from cell to cell. This insight combined with Rule 1 results in the updated rule

$$\phi_{US}(a) = \begin{cases} MS & \text{if } \exists n((MS(n) \lor UC(n)) \land n \in \nu) \\ US & \text{else} \end{cases}$$

4. If the only UC in space is negated, the originally MS becomes unmarked. So a rule is needed that can transition MS back to US. In other words, a rule that notifies the cells when the reason for their marked state disappears. Since this is a CA, this information must be propagated through other cells, or more precisely, the immediate neighborhood. For this purpose, we introduce

the rank R of an MS cell MS_R. The rule states that at least one neighbor of the MS must itself be a lower ranked MS or a UC. If no neighboring cell satisfies this condition, the cell is unmarked; otherwise, it is set to the lowest rank of the neighborhood+1. Therefore, the rank of each MS represents the smallest number of steps from the next UC, and one can see the rank of a UC itself as 0. Thus, if a UC disappears, the neighboring MS no longer have a valid predecessor and transition to US. The neighboring cells of these freshly unmarked cells also have no valid predecessors and are unmarked too, and so on. However, there may still be cases where oscillatory loops are formed by MS and US, as shown in the following Fig. 2.

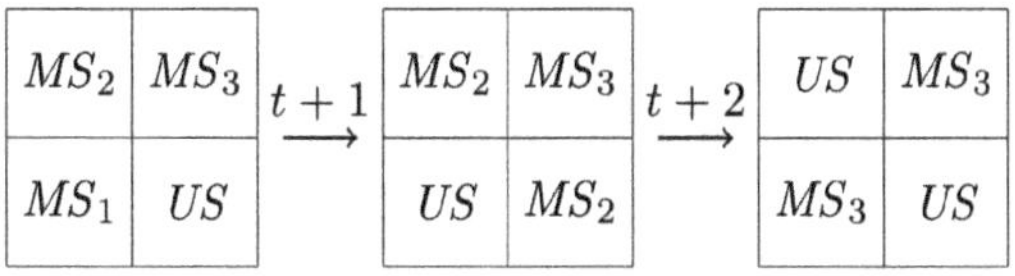

MS_2	MS_3
MS_1	US

$\xrightarrow{t+1}$

MS_2	MS_3
US	MS_2

$\xrightarrow{t+2}$

US	MS_3
MS_3	US

Figure 2: Illustration of how oscillating loops can be formed.

In order to break such loops (remember that the cells only know themselves and their neighbors), a way is needed to make them resistant to a new marking for a while. We define the frozenness F of the Unmarked Space US_F. A MS that changes to US is frozen for F iterations until it may, in US_0, change its state back to marked again. Each iteration decreases the frozenness by 1. It can be seen that with a starting frozenness of 3, the above loop is stopped. However, larger loops are still possible. Larger start values for frozenness make the system less susceptible to such loops, but the responsiveness to change is impaired. A frozenness towards infinity makes the system immune to such loops but also to valid changes. A value of 20 produced good results in the experiments. This gives us the following final rules for US and MS:

$$\phi_{MS_r}(a) = \begin{cases} MS_{\min(R(\nu))+1} & \text{if } \exists n((MS_{x<r}(n) \vee UC(n)) \wedge n \in \nu) \\ US_F & \text{else} \end{cases} \quad (R1)$$

$$\phi_{US_f}(a) = \begin{cases} US_{f-1} & \text{if } f > 0 \\ MS_{\min(R(\nu))+1} & \text{if } f = 0 \wedge \exists n((MS_r(n) \vee UC(n)) \wedge n \in \nu) \\ US_0 & \text{else} \end{cases} \quad (R2)$$

5. Next, rules for the crosses are needed. If the inside of a cross is unmarked, the cross itself should be unmarked and the unmarked space outside should be marked. However, the outside of a cross shall not influence the inside. It becomes clear that a distinction between inside and outside is necessary, otherwise oscillations would occur at the boundary. Also, it is not clear whether the right, left, top, or bottom side of an unmarked cross should be marked. The solution to always only mark the left side, for example, would not allow surfaces to be bounded, since the right side of the boundary would mark the left inner surface. A single cell cannot prevent this. We need to make a distinction beforehand, between the inside and outside. Consequently, the operation is turned into an asymmetrical one, so that the inside and the outside are handled differently and the information can propagate only through one direction. Consequently, we need two additional types of marked and unmarked crosses, inner and outer crosses (IC, OC). IC should be marked if there is a marked cell in the neighborhood. OC should be marked if there is a marked inner cross in the neigborhood. All other cells do not react to the presence of IC. These two rules ensure that the outer space does not influence the inner space across the boundary. One might say that the distinction between the inside and the outside must already have been preceded by a distinction between the inside and the outside.

$$\nu_{IC}(a) = \begin{cases} MIC & \text{if } \exists n (MS(n) \wedge n \in \nu) \\ UIC & \text{else} \end{cases} \quad \text{(R3)}$$

$$\nu_{OC}(a) = \begin{cases} MOC & \text{if } \exists n (MIC(n) \wedge n \in \nu) \\ UOC & \text{else} \end{cases} \quad \text{(R4)}$$

By these rules, there is the limitation for the cross creation that one OC must always be directly connected with an IC. This means that the wall thickness of each cross is exactly two cells thick. In Fig. 3, the smallest valid cross is represented. With this, all the necessary rules required for a viable CA were derived. R1, R2 deal with space, and R3, R4 with crosses. The edge cases that are often solved in CA using a toroidal world can be treated here simply, and in the manner of LoF, by handling the CA boundaries as an unwritten cross.

Finally, by specific choice of the numerical representation for the alphabet S, one can encode all necessary states efficiently with one variable per cell (see Fig. 4).

MS_0	UOC	UOC	UOC	MS_0
UOC	UIC	UIC	UIC	UOC
UOC	UIC	US_x	UIC	UOC
UOC	UIC	UIC	UIC	UOC
MS_0	UOC	UOC	UOC	MS_0

Figure 3: Representation of the simplest valid unmarked cross

This gives us rules that can operate efficiently on numerical values. This final transfer function ϕ is listed below in pseudocode (R).

US_F	$US_{...}$	US_1	US_0	MIC	UIC	MOC	UOC	MS_1	MS_2	$MS_{...}$	MS_r
$-4-F$	...	-5	-4	-3	-2	-1	0	1	2	...	r

Figure 4: Numerical representation of all cell states.

$$\phi(a) = \begin{cases} a+1 & \text{if } a < -4 \\ min(\nu \geq 0) + 1 & \text{if } a = -4 \wedge any(\nu \geq 0) \\ -2 - int(any(\nu \geq 1)) & \text{if } -3 \geq a \geq -2 \\ -int(any(\nu = -3)) & \text{if } -1 \geq a \geq 0 \\ min(\nu \geq 0) + 1 & \text{if } 0 < a \wedge any(0 \leq \nu < a) \\ -4 - F & \text{else} \end{cases} \quad (R)$$

4 Experiments

4.1 Standard case

The derived rules were implemented and tests were performed. The most important cases of LoF are presented below. It should be mentioned that the CA does not function reasonably until distinctions are drawn. To use Niklas Luhmann's words, (Luhmann 2006, 43) "[d]raw a distinction, otherwise nothing will happen at all. If you are not ready to distinguish, nothing at all is going to take place." So in the sense of Spencer-Brown and Luhmann, a user must interact with the CA and "[d]raw a distinction" (Spencer-Brown 1979, 3).

Due to the fact that the cells in the CA can only transmit information per tick by their own change of state, the speed of information transmission is finite. This case can be seen in Figures 5, 6, 7. In Fig. 5, the space is marked with limited speed. In Fig. 6, the negation only becomes active with a delay when the markedness reaches the outer cross. In Fig. 7, there is even a shockwave/ripple resulting from it, which spreads through outer space. This is due to the fact that the cells surrounded by the new cross change their state only gradually.

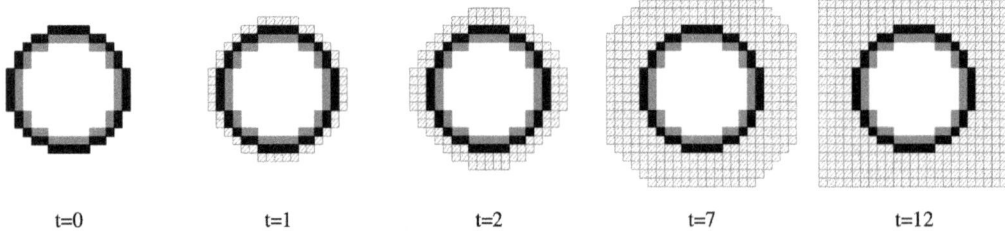

Figure 5: A cross marks an unmarked space. The necessary preceding distinction for the distinction between IC and OC is visible. t indicates the number of elapsed ticks.

As already mentioned, the finite transmission speed is compatible with George Spencer-Brown's thoughts (Spencer-Brown 1979, 59). However, such ripples do not produce intended situations with respect to the re-entry, as we will see. It would be interesting to see to what extent such ripples would also have to emerge in Spencer-Brown's original thought. The unstable states between the initial and final state, however, can be ignored in the normal case without re-entry, since, with

Cellular Laws of Form

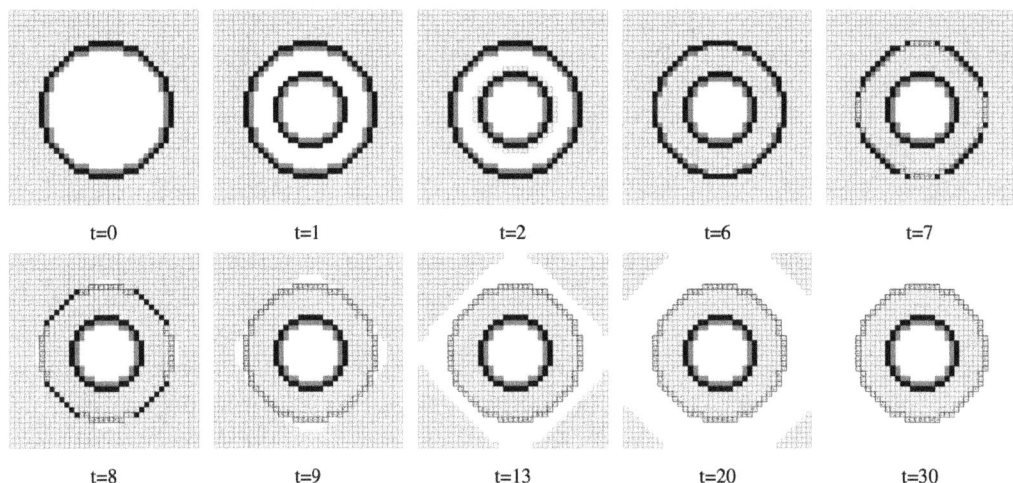

Figure 6: An additional cross negates the first one in $t = 1$. This creates two new spaces, one marked and one unmarked. The dotted and lighter parts of cross cells represent that these cells are in MIC and MOC states, respectively. After 30 iterations, the CA reaches a stable state satisfying I2.

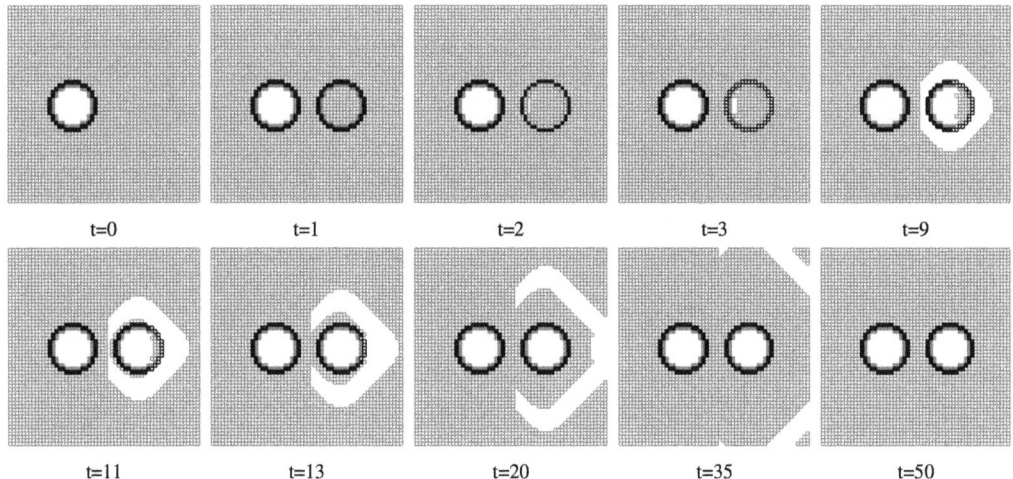

Figure 7: At $t = 1$ an additional cross appears. It can be seen how the delayed retraction of the marked space inside the new cross creates a shock wave of unmarkedness outside. After 50 iterations, the CA reaches a stable state satisfying I1.

sufficiently fast ticks, the change can act instantaneously for the user. Thus, it has been demonstrated that the two axioms can be processed by means of CA.

The next step is to show the laws still hold true for compositions of crosses, which are equivalent to logical operations. For the logical operations, we insert variables as operands. A variable can represent the marked or unmarked state, or, equivalently, an unmarked or marked cross. We interpret the marked state (thus unmarked cross) as true and an unmarked state as false. The NOT operation was represented by means of the recross (Fig. 6). Table 1 presents some basic logical operations. From the operations presented, it is shown that the CA is functionally complete in the same manner as LoF. With the NOT, AND, and OR conjunction, every disjunctive normal form can be formed; alternatively, NAND is sufficient for this (Wernick 1942). Thus, any truth table can be implemented using this CA. Additionally, the operations are relatively visually simple and comprehensible in contrast to normal Boolean logic. Direct relationships between logical functions can be visually inferred. A NOR is simply a negated OR. An XOR is just a negated/recrossed equality. An AND is a negated/recrossed NAND (I2 deletes the two outermost crosses).

As an example that any truth table can be implemented, an implementation of a full adder is presented below. Full adders can add two bits x, y and one bit representing the incoming overflow c_{in} and return a result S with an additional c_{out} as an overflow. A full adder consists of two parts: One part that takes care of the carry and one that takes care of the output. The output can be understood as a parity gate consisting of two successive XOR gates $P = x \oplus y \oplus c_{\text{in}}$. The variable c_{out} is true if at least two true values are present (Ndjountche 2016, 119–120). To implement this full adder in the CA, the corresponding equations must first be transferred into the system of LoF.

$$S = x \oplus y \oplus c_{\text{in}} = \bar{x}\,\bar{y}\,c_{\text{in}} \lor \bar{x}\,y\,\bar{c}_{\text{in}} \lor x\,\bar{y}\,\bar{c}_{\text{in}} \lor x\,y\,c_{\text{in}}$$

The equation can be directly transformed into LoF and then simplified.

$$\begin{aligned} S &= \overline{\overline{\overline{x}\,\overline{y}\,\overline{c_{\text{in}}}}\; \overline{\overline{x}\,y\,\overline{c_{\text{in}}}}\; \overline{x\,\overline{y}\,\overline{c_{\text{in}}}}\; \overline{x\,y\,c_{\text{in}}}} \\ &= \overline{\overline{x\,y\,\overline{c_{\text{in}}}}\; \overline{x\,\overline{y}\,c_{\text{in}}}\; \overline{\overline{x}\,y\,c_{\text{in}}}\; \overline{\overline{x}\,\overline{y}\,c_{\text{in}}}} \end{aligned}$$

Cellular Laws of Form

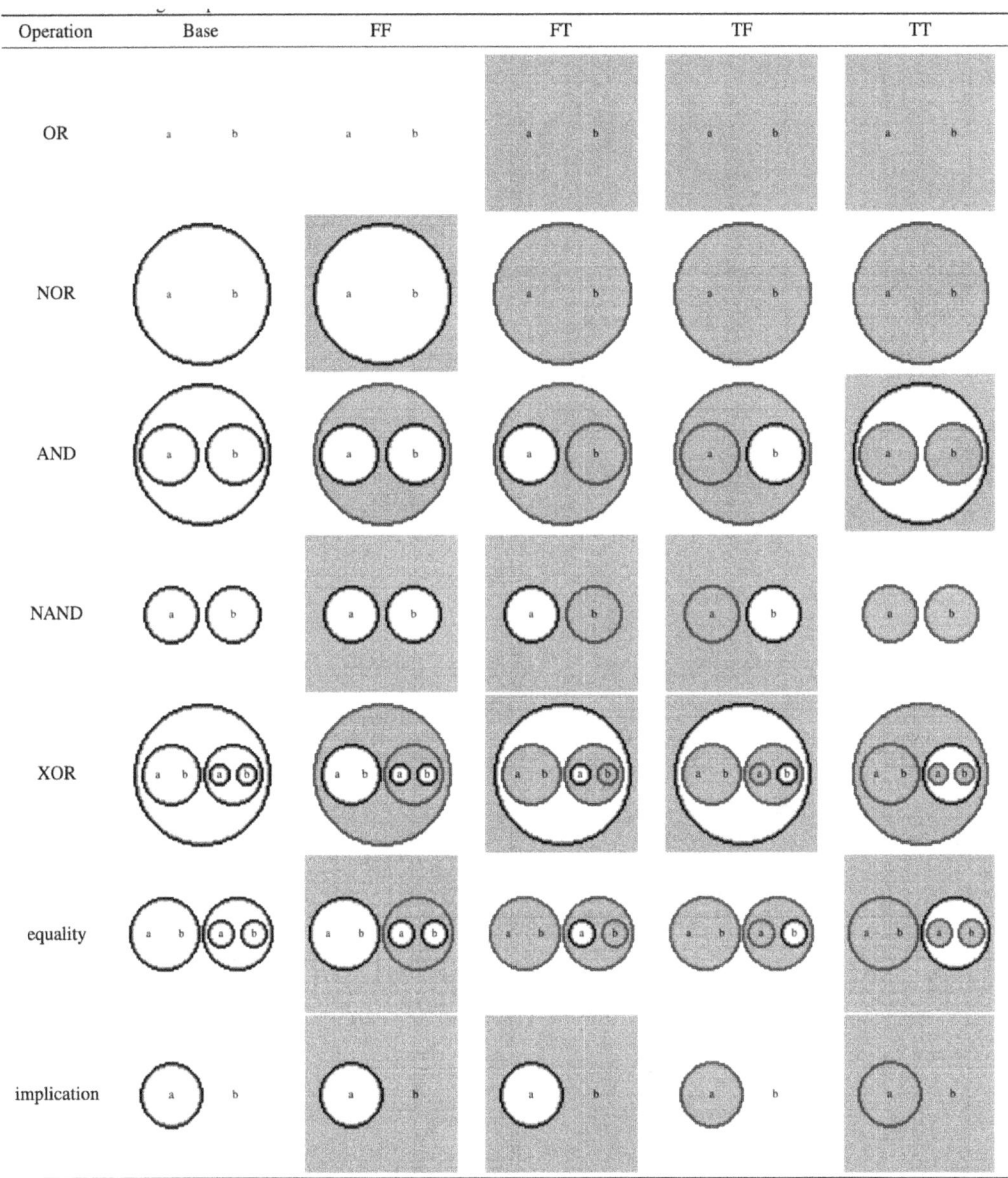

Table 1: Various logical operations are reproduced with crosses. a, b are placeholders that can accept True (Marked) and False (Unmarked) as truth values. The base case is at $t = 0$, all other cases are taken at $t = 150$. The outermost marked space can be seen as the result of the logical operation.

The same can now be done for the carry bit c_{out}.

$$c_{\text{out}} = (c_{\text{in}} \wedge (x \oplus y)) \vee (x \wedge y)$$
$$= (\bar{x} \wedge y \wedge c_{\text{in}}) \vee (x \wedge \bar{y} \wedge c_{\text{in}}) \vee (x \wedge y \wedge \bar{c}_{\text{in}}) \vee (x \wedge y \wedge c_{\text{in}})$$

$$
\begin{aligned}
c_{\text{out}} &= \overline{\overline{x}\,|\,\overline{y}\,|\,\overline{c}\,|}\; \overline{\overline{x}\,|\,y\,|\,\overline{c}\,|}\; \overline{\overline{x}\,|\,y\,|\,c\,|}\; \overline{\overline{x}\,|\,y\,|\,c\,|} \\
&= \overline{x\,y\,|\,\overline{c}\,|}\; \overline{x\,|\,y\,\overline{c}\,|}\; \overline{x\,|\,y\,c\,|}\; \overline{\overline{x}\,|\,y\,c\,|} && (C1) \\
&= \overline{x\,\overline{y}\,|\,\overline{c}\,|}\; \overline{x\,|\,y\,\overline{c}\,|}\; \overline{\overline{x}\,|\,\overline{y}\,|\,c\,|}\; \overline{\overline{x}\,|\,\overline{y}\,|\,c\,|} && (C1) \\
&= \overline{x\,\overline{y}\,|\,\overline{c}\,|}\; \overline{x\,|\,y\,\overline{c}\,|}\; \overline{\overline{x}\,|\,\overline{y}\,|} && (C6) \\
&= \overline{x\,\overline{y}\,|\,\overline{c}\,|\,\overline{x}\,|\,y\,\overline{c}\,|}\; \overline{\overline{x}\,|\,\overline{y}\,|} && (C1) \\
&= \overline{\overline{|\,x\,y\,|\,\overline{c}\,|\,\overline{x}\,|\,y\,\overline{c}\,|}}\; \overline{\overline{x}\,|\,\overline{y}\,|} && (C1) \\
&= \overline{\overline{|\,x\,\overline{y}\,|\,\overline{x}\,|\,y\,|\,|\,\overline{c}\,|\,\overline{|}}}\; \overline{\overline{x}\,|\,\overline{y}\,|} && (C8) \\
&= \overline{x\,\overline{y}\,|\,\overline{x}\,|\,y\,|\,|\,\overline{c}\,|}\; \overline{\overline{x}\,|\,\overline{y}\,|} && (C1) \\
&= \overline{x\,\overline{y}\,|\,x\,|\,x\,\overline{y}\,|\,y\,|\,|\,\overline{c}\,|}\; \overline{\overline{x}\,|\,\overline{y}\,|} && (J2) \\
&= \overline{x\,\overline{y}\,|\,x\,|\,x\,\overline{y}\,|\,y\,|\,\overline{c}\,|}\; \overline{\overline{x}\,|\,\overline{y}\,|} && (C1) \\
&= \overline{\overline{y}\,|\,x\,|\,x\,\overline{y}\,|\,y\,|\,\overline{c}\,|}\; \overline{\overline{x}\,|\,\overline{y}\,|} && (C2) \\
&= \overline{\overline{y}\,|\,x\,|\,\overline{x}\,|\,y\,|\,\overline{c}\,|}\; \overline{\overline{x}\,|\,\overline{y}\,|} && (C2) \\
&= \overline{\overline{x}\,|\,\overline{y}\,|\,xy\,|\,\overline{c}\,|}\; \overline{\overline{x}\,|\,\overline{y}\,|} && (C1) \\
&= \overline{xy\,|\,\overline{c}\,|}\; \overline{\overline{x}\,|\,\overline{y}\,|} && (C2) \\
&= \overline{\overline{x}\,|\,\overline{y}\,|\,|\,\overline{c}\,|}\; \overline{\overline{x}\,|\,\overline{y}\,|} && (C1)
\end{aligned}
$$

Cellular Laws of Form

$$= \overline{\overline{x\,|\,c|}\,\overline{|y|\,c|}}\,|\,\overline{x|\,y|}\,| \qquad (J2)$$

$$= \overline{x\,y|}\,|\,\overline{\overline{x|\,y|}\,|\,c|}\,|\,\overline{x|\,y|}\,| \qquad (C1)$$

$$= \overline{x|\,c|}\,|\,\overline{y|\,c|}\,|\,\overline{x|\,y|}\,| \qquad (C1)$$

It can be seen that theorems of LoF have been applied to simplify the equations, e.g. C1 ($\overline{\overline{a|}} = a$) and C6 ($\overline{a|\,b|}\,|\,\overline{a|\,b|} = a$).

The formulas were transferred into the CA, and thus, a LoF full adder was reconstructed and simulated. The single circuit and all possible states for C_{out} and S can be seen in Table 2 (provided after the bibliography). It can be seen that the output completely resembles the desired behavior of a full adder. For cases without re-entry, the CA satisfies all the rules presented by Spencer-Brown. Next, the re-entry case is examined.

4.2 Re-entry

As mentioned on page 67, one of the most staggering concepts of LoF is the so-called re-entry. It can be shown that the algebraic expression $f = \overline{a|\,b|}$ can be validly transformed into the following infinitely expandable expression through the use of the two axioms and the derived theorems (Spencer-Brown 1979, 55).

$$f = \overline{\overline{\overline{a|\,b|}\,a|\,b|}\,a|\,b|}$$

That can be transferred into this expression

$$f = \overline{f\,a|\,b|}$$

This self-referential expression leads to the re-entry. An expression which is the same at each level as at the previous level (ibid., 56). This expression introduces into space the concept of time (ibid., 58). The re-entry can be used to create flip-flops and thus introduce memory into circuits (ibid., 61). Varela introduced the idea that the re-entry can produce a third state that is neither reducible to the marked nor unmarked state, the so-called autonomous state (Varela 1975, 6–7).

$$\overline{\dots\overline{|}} = \rfloor$$
$$\rfloor = \overline{\dots\overline{|}} = \rfloor$$
$$\overline{\rfloor} = \overline{|}$$

It can be seen that the distinction or negation of the autonomous state has no effect. In this, this state differs from the two previous states. However, the autonomous state is extinguished by a distinction in the same space (Varela 1975, 7).

To mimic the effect of this quasi-wormhole between spaces, we introduce labels. A label, here p, no longer represents a variable but denotes the space. Consequently, a label at the outermost level of distinction is equivalent to the entire expression. Thus, a label represents a connection if it appears in multiple spaces. If a label stands in a marked space, this markedness applies to all positions of this label. The simplest case of re-entry is what Spencer-Brown calls an oscillator $p = \overline{p|}$ (Spencer-Brown 1979, 60–61). This can be replicated in the CA as shown in Fig. 8. Just as Spencer-Brown described, periodic oscillations are produced.

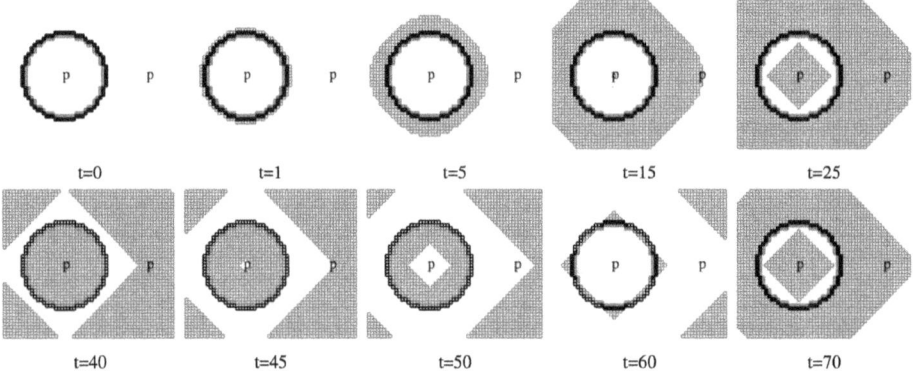

Figure 8: The oscillator function from LoF. Compared to the previous figures, p does not denote a variable but the space and thus represents a link between inside and outside, i.e. a tunnel. The different iterations present the permanent oscillations resulting from this circuit.

In comparison, as seen in Fig. 9, the clock edges are not hard and do not represent a square wave but are soft and more sinusoidal when observing the percentage ratio of the marked space. This is again due to the fact that the information in the CA cannot be transmitted instantaneously. However, this is in accordance with Spencer-Brown's thought experiment that the transmission speed of space is constant: "If we consider the speed at which the representation of value travels through the space of the expression to be constant, then the frequency of its oscillation is determined by the length of the tunnel. Alternatively, if we consider this length to be constant, then the frequency of the oscillation is determined by the speed of its transmission through space" (Spencer-Brown 1979, 59).

If one wants to get rectangular oscillations and approach the original behavior of LoF, it is necessary to have enough time for the spaces to stabilize completely. This can be achieved by allowing the update to pass through the labels only every X ticks. So sampling can be seen as a form of a signal clock and the connection of the re-entry as a clock-controlled line. Also, we can think of the sampling rate—although not entirely—as a kind of tunnel length. The above oscillation function can be well approximated to a square wave with a label delay of 250 ticks, as seen in Fig. 9. Interestingly, similar overshoots occur as in normal square wave oscillations generated by superimposed oscillations.

Problems arise with the memory function $p = \overline{b \, \overline{ap}\rceil}\rceil$. a and b represent variables/placeholders, while p still represents a spatial label. Simply setting the variables is no longer sufficient to achieve the desired behavior. The delay results in oscillations with any configuration, which cannot be suppressed without further effort (Fig. 10). This can be explained by the fact that the markedness in the inner cross must first propagate in order to maintain the markedness in the unwritten cross. Here one notices that the CA has deficits in the faithful reproduction of the dynamics of LoF. Perhaps, however, this could highlight problems with Spencer-Brown's analogy, but this requires more detailed future investigation.

However, one can get the desired behavior by giving the CA enough time to stabilize before sampling the labels. This means setting the variables to the desired values long enough and making the sampling frequency low enough to only ever sample in a stable state. Another way to deal with this is to apply new changes only when the CA has entered a stable state, hence when two consecutive states are identical. Unstable results closer to LoF are obtained with one-way tunnels (which we do not use for the experiments, but could be implemented even more easily) that allow the signal to propagate in only one direction. The extent to which this problem affects Spencer-Brown's idea of the velocity of space is something that could be explored in more detail.

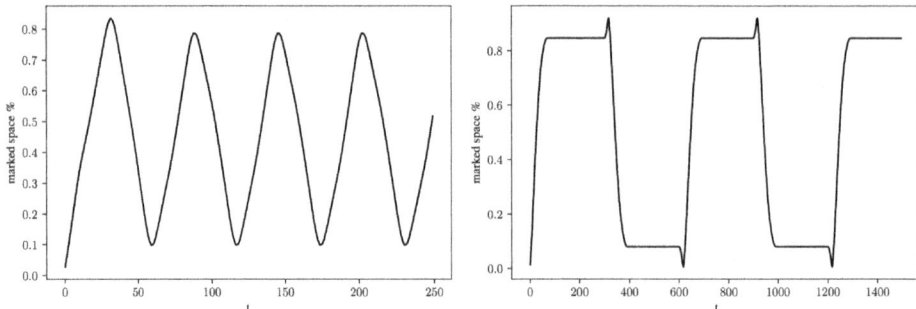

Figure 9: Time courses of an oscillator function. The ordinate represents the percentage of marked space in the CA. On the left, the curve is shown without delayed sampling. The function operates at a uniform frequency. No steep clock edges can be seen. After the first period, the oscillating circuit has settled into a stable waveform. On the right, the curve is shown with delayed sampling. It takes 250 ticks for new changes to be applied from one space to another. The ordinate represents the percentage of marked space in the CA. Compared to the previous variant, rectangular curves are visible. The overshoots at the beginning of each edge are due to additional space being un/marked until the signal reaches the edge of the cross.

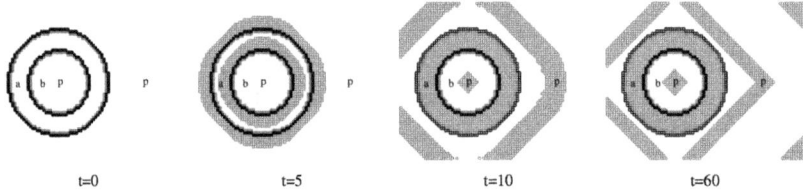

Figure 10: The memory function. a and b represent variables; p the connection between the spaces. It can be seen that the delay of the markedness causes permanent oscillations.

In the case shown in Fig. 11, the labels are updated every 49 ticks. The memory function is indeed represented. The behavior is almost similar to that of an SR flip-flop. The variable a can be used to reset the unwritten cross to the unmarked state (regardless of the value of b). If a is not set and b is set to the marked state, the value of b is stored. After 100 ticks, both variables are set to the unmarked state, it can be observed that this indeed represents the storage behavior and the

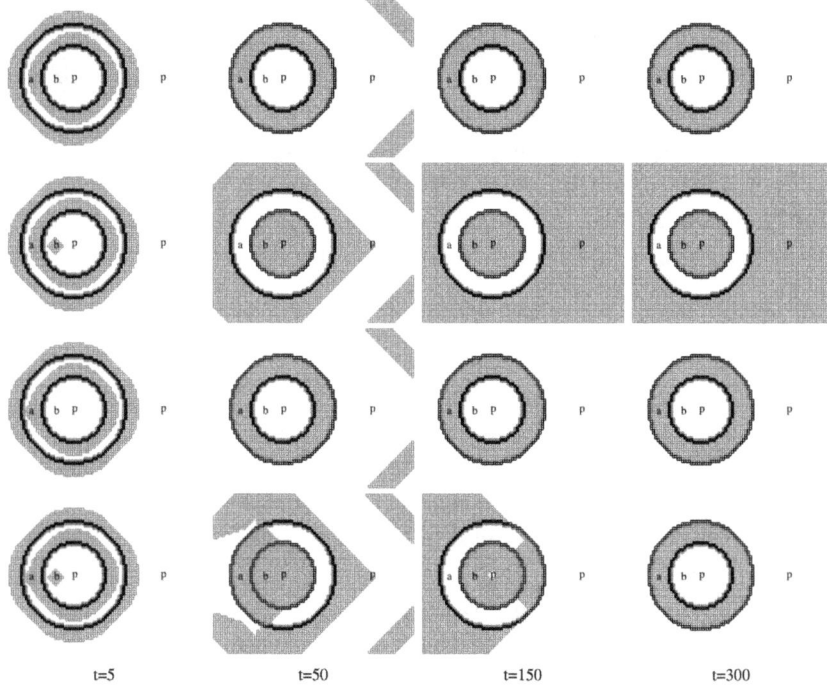

Figure 11: The memory function with delayed sampling. Every 49 ticks the spatial label is updated. The mapping from top to bottom for *ab*: 00, 01, 10, 11. All images were taken from left to right at $t = 5, 50, 150, 300$. After 100 ticks the variables are deactivated. Each occupancy reaches a stable final state. The last mapping (11) takes the longest time to stabilize. At $t = 150$ the unmarkedness sampled in the time step before must first propagate to the outside.

last value remains permanently stored. It has now been shown that re-entry and even the memory function can also be simulated with limitations in a CA.

5 Conclusion

Foremost, it must be said that LoF naturally goes deeper than the two-dimensional representation outlined by Spencer-Brown which has been taken up here.

Nevertheless, the bridge between CA and LoF explored here provides new questions, suggestions, and insights for both fields. A viable CA for LoF was derived. Thus, it has been shown that the dynamics of LoF can be simulated using a CA. Certainly, there are other and also viable rules for LoF state transitions that are to be discovered. Thanks to LoF and its functional completeness, combinatorial circuits like full adders can be realized in the CA. With the help of the controlled re-entry, even memory elements and oscillators are realizable. It would be interesting to investigate other circuits that can be constructed within this CA. It is conceivable that together with the re-entries, complex digital systems such as CPUs could theoretically be created. Likewise, thoughts on the re-entry between CA and LoF would be exciting. Since complex digital systems like CPUs etc. might be rebuilt theoretically, a CA for LoF could be simulated in the CA for LoF—which would bring one back to re-entry.

Probably the most important result of this CA is that Spencer-Brown's thoughts about time, frequency, and velocity can be intuitively simulated and find a plausible representation here. In addition, the CA shows us further insights that can be analyzed in the light of LoF, such as the fact that the border itself must first be differentiated between its interior and exterior sides (here cells) before it can become functional or that creating a cross in a marked space initially creates a shockwave of umarkedness due to the finite speed/inertia of the (un)markedness (Fig. 7). Regarding philosophical implications for LoF, it is interesting to see that these dynamics imagined by Spencer-Brown can emerge from local rules (natural laws, so to speak)—with the remark that these rules are more complicated than the laws of form. The significance of this could be addressed in more detail in the future. The problems with the unstable memory function could perhaps reveal problems with Spencer-Brown's analogy, but this requires more detailed future investigation. The re-entry can probably be realized more naturally with a higher-dimensional cellular automaton, which still needs to be researched. However, the same problems with the oscillations will likely arise and, compared to the tunneling labels, it will not be straightforward to transfer only one state once the CA has stabilized.

References

[1] Das, D. 2012. A survey on cellular automata and its applications. In Krishna, P. V., Babu, M. R., and Ariwa, E., editors, *Global Trends in Computing and*

Communication Systems, 753–762, Berlin, Heidelberg: Springer.

[2] von Foerster, H. 1984. *Observing Systems*. Seaside, CA: Intersystems Publications.

[3] Varela, F. J. and Goguen, J. A. 1978. The arithmetic of closure. *Journal of Cybernetics*, 8(3-4): 291–324.

[4] Gardner, M. 1970. Mathematical games. *Scientific American*, 223(4): 120–123.

[5] Heylighen, F. 1992. Non-rational cognitive processes as changes of distinctions. In van de Vijver, G., editor, *New Perspectives on Cybernetics: Self-Organization, Autonomy and Connectionism*, 77–94. Dordrecht: Springer.

[6] Isaacson, J. and Kauffman, L. H. 2016. Recursive distinctioning. ArXiv preprint arXiv:1606.06965.

[7] Kauffman, L. 2013. Laws of form and topology: Presentation and discussion. *Cybernetics & Human Knowing*, 20.

[8] Luhmann, N. 2006. System as difference. *Organization*, 13(1): 37–57.

[9] Ndjountche, T. 2016. *Digital Electronics 2: Sequential and Arithmetic logic circuits*. Hoboken: Wiley.

[10] Ollinger, N. 2008. Universalities in cellular automata; a (short) survey. *JAC 2008—Journees Automates Cellulaires, Proceedings*.

[11] Orchard, R. A. 1975. On the laws of form. *International Journal of General Systems*, 2(2): 99–106.

[12] Packard, N. H. and Wolfram, S. 1985. Two-dimensional cellular automata. *Journal of Statistical physics*, 38(5): 901–946.

[13] Spencer-Brown, G. 1979. *Laws of Form*. New York: Dutton

[14] Varela, F. J. 1975. A calculus for self-reference. *International Journal of General Systems*, 2: 5–24.

[15] Von Neumann, J. and Burks, A. W. 1966. Theory of self-reproducing automata. *IEEE Transactions on Neural Networks*, 5(1): 3–14.

[16] Watzlawick, P. 1984. *The Invented Reality: How do we know what we believe we know? (Contributions To Constructivism)*. New York: Norton.

[17] Wernick, W. 1942. Complete sets of logical functions. *Transactions of the American Mathematical Society*, 51: 117–132.

[18] Wolfram, S. 2002. *A New Kind of Science*. Champaign, IL.: Wolfram Media.

[19] Zuse, K. 1969. Rechnender Raum (calculating space). *Schriften Zur Datavararbeitung*, 1.

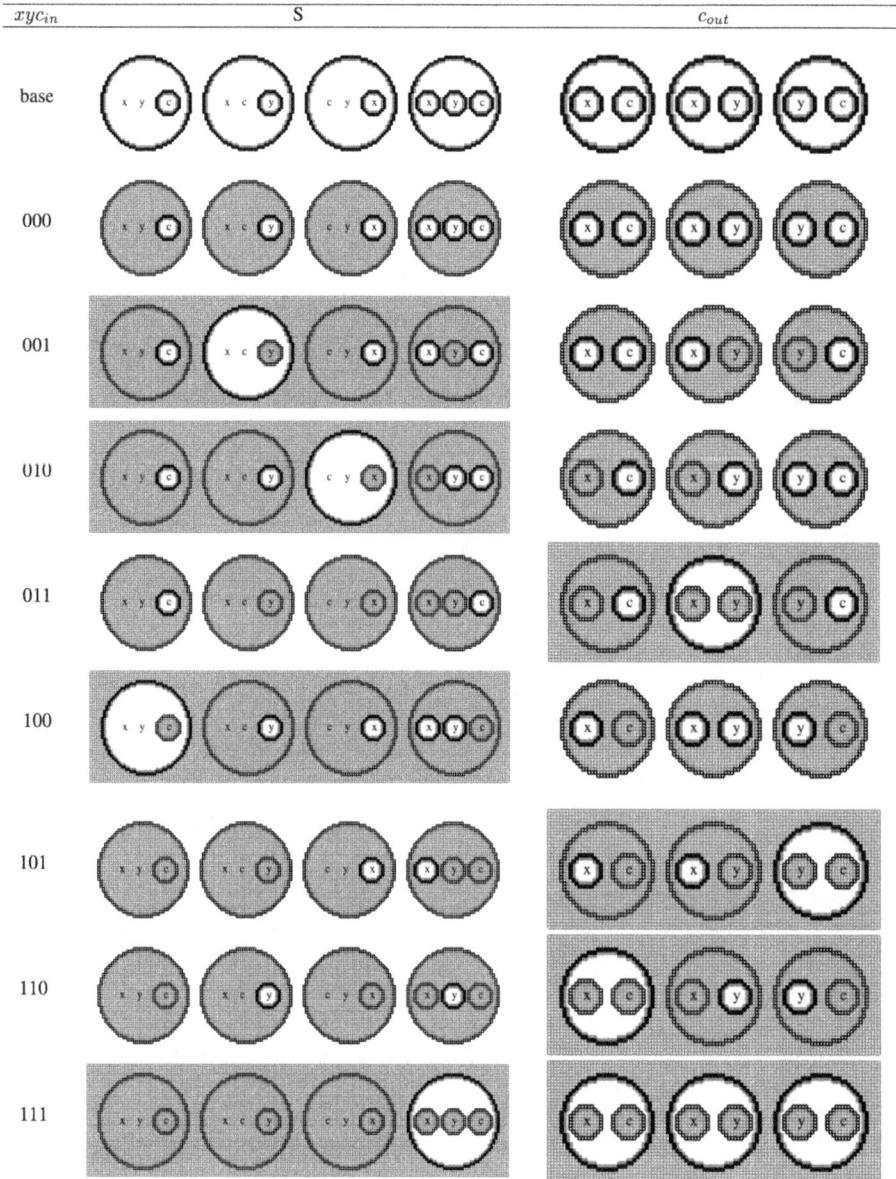

Table 2: A full adder implemented in the CA. All possible cases are shown. x, y and c_in are the inputs. c_out represents the carry and S the result. The base case is at $t = 0$, all other cases are taken at $t = 250$. The outermost marked space can be seen as the result of the logical operation.

Forms of Decision Models in Organizations

Florian Grote
CODE University of Applied Sciences, Berlin
`florian.grote@code.berlin`

Abstract

The social world of communication is built on decisions. Many of these are microscopic operational decisions, made when choosing what to say next in a discussion to keep the conversation going or to use tactics to win an argument. Other decisions might seem big in comparison when it comes to the direction in which a company is developing its strategy. For such decisions, we usually observe an approach to decision making that seems more structured than for the multitude of micro-decisions. This paper is an exploration of these seemingly structured approaches. The exploration takes cues from sociological systems theory and an analytical perspective that focuses on the construction of perceived reality by observations. Thus, the empirical field discussed is that of communication and social phenomena that emerge from it. Decision making is often explored as a process in individuals, rendering it a study field for psychology. While not contesting the rich set of psychological phenomena impacting decision making, this paper discusses primarily the social side of the distributed process needed to make and take decisions.

The author would like to thank the reviewers for their efforts and the valuable suggestions for revisions.

1 Decisions and Observations

Common to decisions of any scope is that they are made in a world of undecidable paradoxes, as otherwise, there would be nothing to decide on (Gödel 1931; von Foerster 2003). The principal function of a decision is to choose among a range of options in a situation where no preference for any of the options has been established yet. To get to a decision, some form of preference model must be established which allows a decider to choose one of the options. This is a complex endeavour, as we will explore.

Dirk Baecker (2019) has pointed out that any decision could also not be taken, thus creating a sort of null-hypothesis as the baseline to consider other options against. Any choice that does get made thus contains both the negation of not making it and the negation of choosing the different options. This makes decisions consequential, as they define a path of action that might otherwise not have been taken. This is consequential both for the social situation they impact and for the decider who will be held accountable for the outcome of the decision. Niklas Luhmann (2018) has described decisions as forms of social observations, where the decision is not just about making a choice among a range of independent options, but where the decision also manifests itself in the specific observation of the world as a situation where the choice has been (had to be) made. Thus, the decision unfolds in the temporal dimension: The decision itself is observable as an action taken, namely the action of continuing to operate with the choice having been made. Thus, the decision comes into the world as a new observation: A choice has been made, and from now on, all potential options have to be observed in the context of the choice (for this action) having been made. The decision makes a distinction between a world before and a world after the choice: Before, options were tangled up in undecidable paradoxes, and afterwards, options have to be observed in the context of the choice. To make a decision out of the undecidable, an observation is necessary.

In addition, the observation introduces a communicational address to the situation. The decision can be attributed to a decider who is held accountable for the new observation of the world after the choice has been made. Organizations have found numerous ways to define decision attribution and deal with the politics surrounding it.

Part of the new observation after the choice has been made is often a narrative surrounding the choice, embedding it into a sequence of described events that make the choice seem inevitable. This is known as post rationalization and has been described as making choice seem as the best (or only) fit in the context of expectations before the choice. Essentially, this is about making it seem as though the situation before the choice had not been undecidable. Luhmann (Luhmann 2018, 104) has described this as an observation of its own decision situation, distinct from the flow of communication up to the choice event and in its own temporality, a re-entry of the before and after of the decision into the narrated before and after of the decision.

This re-entry enables the narrative of bounded rationality (Simon 1972) for an organization, constructing for itself and its stakeholders a heuristic model for decision making. This paper argues that such heuristics are already the outcome of the organization constructing its own reality by selectively "enacting" (Weick 2000; 2010) complexity in its environment.

2 Forms of Communication

For the formal analysis of communicational constellations, we use George Spencer-Brown's calculus of indications, as expressed in the book *Laws of Form* (Spencer-Brown 1969; 2008). From it, we take the fundamental constructor of the distinction and apply it to social situations, where it separates and thereby connects different areas of meaning in communication. The distinction:

$\overline{\phantom{\rule{1em}{0ex}}}$

In social situations, we take the distinction to create a separation in an area of meaning, constructing a form with an operational inside that is taken to be separate from everything it is not. This "everything it is not" can then be explicated as a specific outside, giving the form a relational focus—we are observing the operational relation between areas of meaning a *and* b*, instead of simply area of meaning* a *as separate from everything it is not.*

$\overline{a}\,b$

Applied in the calculus of indications (Spencer-Brown 2008, 4), the distinction can then be repeated as a call:

or it can be crossed, moving from one area of meaning to the other:

Additionally, the distinction can also be re-entered into itself (Spencer-Brown 2008, 45):

For communication, this relates to a situation where the distinction between two areas of meaning (a and b) is used in the operation of communication within (in our example) area of meaning a itself.

Area of meaning a then does not simply operate as a, but instead operates as the (socially) cognisant area of meaning a as distinct specifically from area of meaning b. This will be demonstrated in examples below.

Social situations can be topics of communication because participants can give descriptions of how they observed them. Thus, the form arrangements created out of the form analysis of social situations are intended to represent descriptions of observations (cf. Grote 2014). The confluence of a narrated before and after as well as the intended interpretation of the choice as inevitable in these situations lend themselves well to the construction of causality. In *The Book of Why*, Judea Pearl (Pearl and Mackenzie 2018, 158–162) describes causality as the construction of listeners, where the observation draws a distinction of which area of meaning is listening to which other area of meaning. The metaphor of listening used here is one of active determination. The observation needs to actively determine which area of meaning is listening to which other area of meaning in order to construct a causality from correlation. The example Judea Pearl gives (Pearl and Mackenzie 2018, 15) is that of the rooster and the sun. Does the sun rise because the rooster crows, or does the rooster crow because the sun rises? Which area of meaning—sun or rooster—took the decision to let its actions be determined by the other? Was this even a decision? Even though most observers might construct the rooster as listening to the sun rising, this relation itself is not available to communication.

Thus, we can only separate the areas of meaning as observers who have knowledge of both sunrise and rooster:

rooster crows = $\overline{\text{rooster} \mid \text{sunrise} \mid}$.

3 First Re-entry

This is different in cases where decision making takes into account actions by other participants in communication. Here, a decision is a special form of a social observation, and the observation of a decision is another such form. Both forms are subject to the double contingency (Luhmann 1995, 103) of communication, meaning that the decider has to take into account that their attribution of the decision to specific actions will be observed by those who carried out these actions and *vice versa*. Take for example the case of the traditional 4-eyes principle in organizational decision making, where another proverbial "set of eyes" is required to observe the decision making. The decider and the observer enter into a state of mutual contingency, which is by design to exclude certain decisions the decider might otherwise take if they could do so in a seemingly unobserved, unchecked state. The decision will therefore be influenced—albeit not determined—by the actions of the observer. However, observing the decision-making process by the decider, the observer might also decide to change their actions based on what they observe. This is the reflection part in the double contingency, where the situation oscillates and both participants switch roles. An application of this can be found in crew resource management (CRM) in aviation (FAA 2017, 6–3; Hagen, Lei, and Shahal 2019): In the cockpit of commercial airliners, there are (at the time of writing) still two pilots, one with the designation "pilot flying" and the other with the designation "pilot monitoring". As crew resource management is all about formalizing communication to prevent misunderstandings that could lead to dangerous consequences, these roles are well-specified and the switch of roles—which can in principle happen any time—is also clearly formalized. From the perspective of an organization (which includes the crew of an airplane), as each area of meaning works with observations of the other observing itself, the question is whether it can continue to operate as it did before and take its own decisions based in the existing social situation or whether something fundamental needs to change in the social situation of decision making. With Luhmann (2018, 180), we call these areas of meaning that have to be taken into account when making decisions "premises". As Luhmann points out, premises are themselves decisions taken in their own social situations of interactions and other premises. In theory, the organization then maintains its focus on organizing premises in such a way that it optimizes for its own survival (cf. for leadership and organizational survival: Richter and Groth 2023; Luhmann 2018, 180), while the usual complexity of social situations makes full optimization unlikely in actuality.

decision interaction = pilot flying | pilot monitoring .

Form of decision interaction: Considering the operation of the pilot monitoring, can the pilot flying continue to operate as before?

The first re-entry we introduce into the exploration of decision making is thus the realization by the operational decision taker that they are acting as such within a situation of mutual observation, an interaction. Nevertheless, it is not just the pilot flying adjusting their actions based on the observation that they are being observed by the pilot monitoring, but also the pilot monitoring taking decisions to change their behaviour based on the observation of being observed by the pilot flying. The situation of mutual contingency is characterized by there not being clear determinations of what might happen next, even though in the example of an airplane cockpit, clear guidelines for crew resource management exist. These guidelines provide an overarching decision model, but they do not determine what happens in any given situation.

Instead, the development of the situation is influenced by the expectations the pilot flying and the pilot monitoring have toward each other. These expectations are typically informed by their experience, the role, and their previous experience with each other on a personal level. Each pilot will have expectations for what the other might do next, but there is no guarantee this is going to happen. Thus, the description of the unfolding of the situation is contingent from both pilots' perspectives. Luhmann has described such a situation of a distinction with mutual contingency for both sides and oscillating perspectives the "unity of the difference" (Luhmann 1995, 4), pointing out that no element of the social form can be observed separately—by an outside observer—if the function of the form is to be communicated. For the decision, this means that the selection of options is not determined by any one observer but framed (in the sense described by Kahneman 2013, 86) by the situation of mutual contingency with assumed expectations.

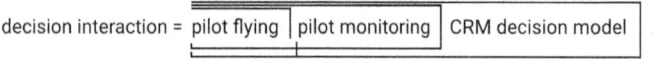

Form of decision interaction: Considering the knowledge of the CRM decision model, and the pilot monitoring's operation and knowledge of the CRM decision model, can the pilot flying continue to operate as before?

The second re-entry in the exploration of decision making makes available the framing of a decision model for the operational realm of the decision taker and their counterpart in the interaction situation.

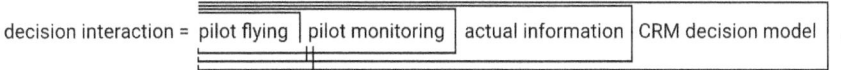

Form of decision interaction: Considering the knowledge of the CRM decision model, the actual information gathered, and the pilot monitoring's operation and knowledge of the CRM decision model as well as the actual information gathered, can the pilot flying continue to operate as before?

Between the decision model and the operational realms, information gathered in the actual situation interferes. In the case of piloting an airplane, this is information from technological systems of the plane, radio communications, and visual information from outside the windows, for example. How the crew deals with this information is also influenced by the CRM decision model. We will come back to this type of form arrangement later.

4 From Action to Observation

When looking at situations like the one described above, we can analyse the flow of communication, but we will ultimately be interested in the actions resulting from decisions taken. In Luhmann's theory (Luhmann 1995), communication is the process of autopoietic operation in the social realm, so in a situation of double contingency, where all participants in the communication observe themselves being observed by others, who react to themselves in ways framed by mutual expectations. This is how meaning is established in communication. Action is the broader course of how communication unfolds across functional boundaries, itself once again observed in communication (Luhmann 1995, 59). It becomes the narrative of how people, things and behaviours are constructed out of the underlying operational process of communication. An analysis of communication and the construction of action such as this one is itself such a construction, thus constituting a re-entry of the process into itself. We can then analyze such constructions as accounts of observations of action. Decisions are embedded into both the operational process of communication, into the construction of actions, and the observation of decision making and its impact. On all levels, continuity of action needs to be achieved by selecting among a variety of options. Decision making thus can be described as being embedded in a network of other decisions, which act in shaping the prerequisites for the decision, themselves having gone through the process of being shaped by other decisions (cf. Baecker 2022, 12). In the social flow of communication, a specific direction for the continuity of social action is chosen from the range of potential options and a new narrative—a new set of actions leading up to the choice—is constructed and made available for observation. The communication address of a decider is often an important part of such a decision narrative. For a decision, an observation of the present is chosen from observable complexity, a narrative of the past is chosen from the pool of plausible histories, options for the decision to consider are chosen from the horizon of possibilities, a decider is appointed from the pool of available addresses in communication, the plan to take the decision is constructed from the flow of communication which could also create other plans, forecasts for outcomes of choices are created with knowledge of uncertain futures, resources are allocated toward a specific set of actions and not to others, and costs incurred because of these actions are considered to be acceptable or not. In addition, all areas of meaning mentioned here see that which they are observed as distinct from re-entered into themselves, resulting in several distinctions having to be taken into account in their operation. With Luhmann (2018, 161) and Baecker (2019), we thus have cases of reflexive-rational decisions: Reflexive in that they observe the core distinctions out of which they are formed and rational in the sense that they create a present that is constructed out of a choice taking the restrictions of its premises into account—or deliberately taking the risk to negate some or all of its premises.

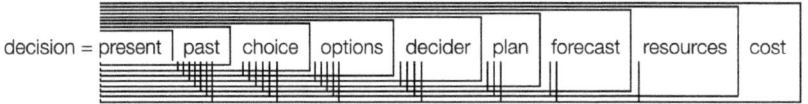

This form of the decision condenses the complex dependencies between areas of meaning by visualizing the distinctions that have to be taken into account for their operation via re-entries. We can approximate the communicational operationalization of these dependencies by setting a focus on the individual areas of meaning included in the form and asking the defining question mentioned above: Considering the dependencies, can this area of meaning continue to operate as before, or is a decision involving other areas of meaning needed?

Forms of Decision Models in Organizations

Form of the decision:

Considering the potential cost, can we continue allocating resources, knowing that it will be taken into account in forecasting, planning, choosing a decider, selecting options, enabling a choice, selecting a narrative of the past, and ultimately constructing a present?

Considering the resources available, taking into account the cost, can we build a promising forecast, knowing that it will be taken into account in planning, choosing a decider, selecting options, enabling a choice, selecting a narrative of the past, and ultimately constructing a present?

Considering our forecast, taking into account resources and cost, can we continue with the plan, knowing that it will be taken into account in choosing a decider, selecting options, enabling a choice, selecting a narrative of the past, and ultimately constructing a present?

Considering the plan, taking into account the forecast, resources, and cost, can we continue with our choice of the decider, knowing that it will be taken into account in selecting options, enabling a choice, selecting a narrative of the past, and ultimately constructing a present?

Considering our choice of decider, which takes into account the plan, forecast, resources and cost, can we continue with our pool of options, knowing that it will be taken into account in enabling a choice, selecting a narrative of the past, and ultimately constructing a present?

Considering our pool of options, which takes into account the decider, plan, forecast, resources and cost, do we have a viable choice, knowing that it will be taken into account in selecting a narrative of the past, and ultimately constructing a present?

Considering our outlook to a viable choice, which takes into account the options, the decider, plan, forecast, resources and cost, can we rely on our chosen narrative of the past, knowing that it will be taken into account in the construction of the present?

Considering our chosen narrative of the past, which takes into account our outlook to a viable choice, the pool of options, the decider, plan, forecast, resources and cost, can we continue with our construction of the present?

5 Decision Models

The prior decisions for decision making are themselves guided by a diverse range of factors, including heuristics and framing (cf. Kahneman 2013). In the following, we will explore how structured decision making can make use of dedicated models of reality that focus on parts of the form of the decision. The first gain of such a model is the alignment of distinct areas of meaning, for example via the integration of the forecast, plan, resources, costs, and options in one concise scenario, which can have its own narrative and be discussed strategically in an organization or a family, for example. Notably, this includes discussions of failing to execute the plan, exhausting the resources, cost overruns, etc., as Dirk Baecker has pointed out. Similarly, the creation of a decision model might be motivated by aligning a specific selection of historical narratives with the choice of the decider, which might also fail. A secondary gain of the construction of a decision model lies in the availability of this model for further transformations, such as tracking successes and losses or trying to determine productivity. This part of model creation focuses on the social-operational dimension of constructing what we refer to as a decision model.

decision model = decision | model of reality | .

Form of the decision model: Considering our model of reality, can we continue with our construction of a decision?

Another aspect of decision models is the realm of forecasting. Predictions and forecasts are one key motivator for decision making. As Luhmann has pointed out (Luhmann 2018, 134), one of the characterizations of a decision is that it contains observations of the past and expectations about the future which are used to create its present. In decision making, we can observe efforts to connect the past and the future via the construction of a causal link between them in the realm of forecasting (cf. Foley and Khavkin 2019). This link then needs operationalization in the present, which has to be decided on. Work on forecasts is often presented as a quest to find the computationally optimal (in some dimension) choice among possible options. For this, the decision model has to be digitized, meaning that forecasts need to include numerical dimensions sampled from the complexity of reality. Sampling is the process of taking selective readings from a measurement of the state of a continuum, thereby creating a model that reduces the complexity of the continuum (cf. Baecker 2022, 14). Sampling is a process of digitizing, where the input is a continuum and the output is expressed in discrete steps, each having a symbol from a finite set of symbols mapped to it (McFee 2022).

The further approach is to attach metrics to options so that the situation clears up and—according to the interpretative model of the metrics—the choice is not undecidable anymore. In the case of business forecasting, it is presented as a risk-benefit trade-off. The function of this quantification of the options is the exclusion of the qualitative uncertainty from the choice. Of course, the uncertainty does not go away, but it is rather transferred to the decision on which metrics to apply and how to construct and interpret the model for the situation. This would be another constructed narrative around how actions might unfold, and how, in that newly established context, a choice among a selection of options can become decidable. This process is not limited to explicit forecasting in economic or political contexts. Eric Johnson describes the construction of alternatives as the assembling of preferences, where "choice designers" or "architects of choice" (Johnson 2022, 8) make predictions about the choices other might make when presented with a set of alternatives they are designing.

One common approach to making predictions in the realm of dynamic systems is that of growth in the context of limited resources, famously used to model growth of living standards in the 1972 Club of Rome report on the Limits to Growth (Meadows et al. 1972). For this report, as well as the ensuing 30-year update (Meadows and Randers 2013) and the recent Earth4All iteration (Dixson-Decleve et al. 2022), the group of authors and contributors built a "world model," aiming to represent the earth-scale impact relationships between the consumption of resources for living standards, the availability of these resources, and impacts caused by the consumption of resources, such as pollution and climate effects. It uses methods from systems modelling, combining variable factors on growth dynamics (Sterman 2000). For growth modelling, the logistic function is used, where growth is assumed to consume finite resources (Weisstein 2003). Thus, the available resources constitute the so-called carrying capacity for the growing entity.

This approach creates a digital model of how the observed world is expected to change if a decision is implemented. The model is a digitalization and algorithmization of the set of expectations out of which it was created, very similar to the social part of the decision model described above. It is very much a representation of an observer perspective and functions to make specific parts of expectations explicit and available for comparison against different scenarios of how a situation might unfold. Once a model is established, it can be fed different values for how its metrics might develop. The results in the model can then again be sampled (remaining in the digital realm) and used to contextualize the observation of decision options available in the decision situation. In the example of Earth4All

(Dixson-Decleve et al. 2022, 27), the authors propose two scenarios: "Too Little, Too Late," where over-consumption of resources continues to drive society into crisis, and the "Giant Leap" scenario, where fundamental changes are implemented to allow for the sustainable development of an inclusive and equitable civilization. With these scenarios, the authors aim to frame further discussions in decision making, as they visualize which impacts could be expected over time as decisions support one or the other model. For this to work, both scenarios must be communicated together. The "Giant Leap" scenario needs the "Too Little, Too Late" scenario as the stick to serve as a carrot and *vice versa*. This has been described by Luhmann as another characterization of decisions: They do not just combine past and future via constructing a present but also different, potentially adversarial choices (Luhmann 2018, 97).

Creating scenarios with different levels of optimism based on what is known about the audience is standard practice in product planning and business forecasting (Tetlock and Gardner 2015; Foley and Khavkin 2019), augmented by marketing strategies and projections about how the audience and the market it constitutes might develop. Thus, the decision model is expanded with the different scenarios resting on a consistent basis of assumptions for how actions impact resources and *vice versa*.

A scenario-based decision model can thus be described as the unity of the difference between present and possible futures, and expectations as to how the present and the possible futures might be connected.

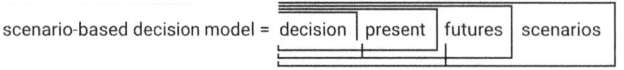

Form of the scenario-based decision model:

Considering our scenarios, can we uphold our narratives of possible futures, knowing they will be taken into account in the construction of the present and the decision?

Considering our narratives of possible futures, taking our scenarios into account, can we continue with our construction of the present, knowing it will be taken into account in the decision?

Considering our construction of the present, taking into account our narratives of possible futures and our scenarios, can we continue with our construction of the decision?

After the decision has been implemented, a scenario-based decision model can be used to compare real-world data with predictions made by the model, especially when a decision directly corresponding to one of its scenarios has been taken. This is where the digital nature of the decision model shows its main function. In cases where detailed forecasts have been created, its use brings the benefit of having an arbitrarily (computationally limited) detailed set of comparison points available for analysis. In the field of physical processes and systems, such a decision model is often a part of a much larger model that encompasses the entire value chain: The "Digital Twin" (Korenhof, Giesbers, and Sanderse 2023).

In business modelling, the system dynamics approach mentioned above is often taken, explicating assumptions and forecasts in dynamic impact models (Sterman 2000). This approach allows the contextualization of actual information on the success of a product with the assumptions from the forecasts as expressed in the digital model. Business decisions can then be made dependent on the relation between the actuals and the forecast. Actuals or actual information reference(s) information gained from the situation observed as the present. For use in analysis, this information is then recorded as a time series, thus making it compatible to a digital model which also operates on time series of data, only that this data is created algorithmically from assumptions.

The explicit use of decision models and their inclusion of dynamic system-level forecasts has shown benefits in organizational applications, such as alignment on strategy and tactics, as well as accountability in implementing decisions as well as constructively critiquing and learning from failures (cf. Rosenzweig 2014). The decision model can be described as a digital, algorithmic, and thereby dynamic connection of reduced complexity between established expectations and observations informed by these expectations once the decision has reached the implementation phase. In that phase, the decision model can be contextualized with information on actual developments. Its function reaches further than simply creating a reaction by the organization to an external stimulus, however intricate the heuristics used in this creation might be (cf. Simon 1972). Instead, the network of sampled expectations and data points from enactments (Weick 2000; 2010) included in the decision model give it a structural function, core to the actions involved in organizing work and stakeholders. This seems inevitable as decision models are not just used to determine an optimal outcome from a bounded rational set of premises, but they actively create their own surplus of meaning (cf. Luhmann 1995, 97) for the organization to process selectively.

6 Social Perspectives on Decision Models

Decision making is a distributed social process. Even decisions perceived as individual include social scenarios as choices. Eric Johnson has called the choices in decisions perceived as individual "assembled preferences" (Johnson 2022, 56), and sometimes these scenarios are deliberately constructed socially, for example in the form of role models or other success stories. Thus, decision models are not built from one individual perspective but rather are the outcome of the interaction of different perspectives. This interaction can take different forms, ranging from explicit, open collaboration on a decision model to parts of the decision model being deliberately hidden from view. In this section, we will start by looking at open collaboration.

As described above, observations of the present, possible futures, and scenarios to connect them must be aligned among those directly involved in the decision and its implementation. An example of such a situation is a product development team planning its next steps. This open collaborative process of decision making is built on the expectation that the perspectives on the decision are communicated transparently. Everyone on the team is observing the structured decision-making process with its artifacts and adds their observer perspective to the discussion (cf. Grote 2023). Notably, this transparency is an expectation, and it functions to structure the social process as such. The process then has to deal with the usual social complexities of misunderstandings, miscommunication, hidden agendas, heuristics, and framings, etc. Transparent, open collaboration can only serve as a goal, while the realities of decision making must be made resilient for a social reality where it is never fully achievable (cf. Luhmann 2017). Nevertheless, the benefits might be worth it, as open interaction (via the surplus of meaning it creates) can facilitate relational accountability and thereby increase the chances of successful implementations of decisions (Richter and Groth 2023, 84).

collaborative decision-making = | decision model | open collaboration | .

Form of collaborative decision making: Considering the state of our open collaboration, can we continue to build our decision model?

While some distributed decision making processes function based on the goal of transparency, others involve perspectives on the decision which are deliberately more implicit or hidden. This is often the case, for example, when sellers are trying to influence the decision making of their prospective buyers via marketing.

Eric Johnson has described dynamic (mostly technological) implementations of decision models as "choice engines" (Johnson 2022, 244), namely as models with alternatives designed to elicit certain desired choices among deciders. This is the case for online platforms and marketplaces, for example, where algorithms are designed to cater to the assumed interests of their users. Such choice engines can be observed in the algorithms of Amazon.com or Facebook, and many similar offerings, as well as the Google search results when they take the user's search history into account. Choice engines dynamically create choice architectures. However, the specific choice architecture for an individual user is usually not observable for that user at the time the user is interacting with the system. Rather, a marketplace like Amazon presents itself just via the choice architecture dynamically generated for any particular user. The impact of these opaque choice engines on decision making have been studied broadly, especially in the economic and political dimensions (e.g. Graham 2018), and their existence recursively triggers discussions of accountability for the organizations which created them.

hidden decision-making = decision model | opaque choice engine .

Form of hidden decision making: Considering the state and success of our opaque choice engine, can we continue to build and operate our decision model? Note that this is an observational perspective on the creators of the hidden decision making process, which is not hidden to them, just as the choice engine is not opaque to them.

7 AI: The "Artificial" Perspective on Decision Models

When decisions are understood as models based on a situation characterized by the multilateral contingency of expectations and actions, the utilization of large language models (LLMs) to build models seems fitting. Typical foundational LLMs are trained on large volumes of textual information, mainly from sources on the open internet, plus published material in the form of books, journals, and news publications (OpenAI 2023; Schaul, Chen, and Tiku 2023; Rogers 2023) With this data as its training base, a significant part of this training material will consist of narratives of decisions taken, discussions of decisions to be made, and constructed causalities in the sense of the listener model by Judea Pearl (Pearl and Mackenzie

2018). Thus, adding the artificial "perspective" of an LLM response to a decision model may have the potential to provide a broad representation of expectations toward the outcome of the choices assembled in the model. Adding LLM-based expectations may thus increase the predictive capabilities of the decision model.

LLM responses are referred to as "artificial" in this case because they do not result from social operation but from interaction with a technical system which itself is not party to social operation, even though it might hold models of functions in social operation. Elena Esposito (2022) has proposed to refer to interaction with AI systems which are able to mimic taking part in communication as "artificial communication". The argument is that AI tools do not succeed because of whatever level of intelligence they might or might not possess but because their output is optimized for signals of participation in communication. Thus, when it comes to observer perspectives, whenever the AI output is capable of signalling descriptions of observation, those observer perspectives should equally be referred to as artificial.

decision-making incl. artificial perspective = $\overline{\text{decision model} \,|\, \text{AI output}}$.

Form of decision making including an artificial perspective: Considering the output we are getting from our AI tools, can we continue to build our decision model?

This raises questions about the social, collaborative aspects of creating decision models. We have covered teams making decisions in open collaborations and the case of sellers like online marketplaces using choice engines to dynamically create choice architectures where the collaboration is more opaque to the eventual action of taking the decision. With LLMs, it becomes possible to add an artificial perspective to the creation of a decision model, including the dynamic choice engines. The question becomes, how does this artificial perspective fit in with the self-observation of the "decider(s)" in the social process of decision making?

8 Experiment

To investigate this question, a group of nine students in a workshop on product-related strategic decision making were asked to build a simple decision model as part of the workshop assignments. All students had running projects with the goal of building a product. They were given a self-developed (by the author) canvas for a basic decision model, which consisted of four fields, each of which held a question to prompt an answer from the participant or a headline to fill with content. The canvas was provided on the collaborative online whiteboard Miro (miro.com). Work on the canvas was done individually but shared in groups of two to three students afterwards. The questions (in one case with hints) and headlines were:

- *What is the goal behind the decision?*
- *Which other premises inform the decision?*
    - *market developments*
    - *available resources*
    - *stakeholder requirements*
    - *technology choices*

Proposed decision:

Expected early indicators that the decision was correct:

Next to the canvas, an example was given as inspiration for the participants. After the participants had created their first own decision model, the participants discussed their outcomes in groups of two or three. Then, there was a round of general feedback with everyone, and two participants volunteered to present their outcomes to the full group. After this, the participants were invited to open a separate web page with an LLM-enabled online program created by the author. This online program consisted of a single-entry field for text, where the participants were asked to enter (via copy and paste) the content of the decision model canvas they had created. Via a button click, this content was then sent to the GPT-4 LLM via the OpenAI application programming interface (API). The tool was designed to not allow the LLM to make direct suggestions of points to add to the decision model but rather to invite further reflection by asking questions relevant to the content. The following is an example of an interaction with the LLM-enabled online program.

Input by the user:

Goal: Increase user engagement rates by tailoring content and interactions to individual user preferences, targeting a 35% rise in personalized interactions within 6 months.

Premises:

- Users are more likely to engage with content that aligns with their individual interests and preferences.
- Current analytics show that a one-size-fits-all approach has led to varying engagement levels.
- AI can analyze user behavior patterns to predict and serve tailored content.

Proposed decision: Integrate an AI-driven personalization engine that analyzes user behavior, historical data, and preferences to curate and present personalized content or product recommendations.

Expected early indicators that the decision was correct:

- Increase in user session durations, indicating more time spent interacting with personalized content.
- Enhanced click-through rates on AI-recommended content or products.
- Positive user feedback on the relevance and quality of content presented to them.
- A decrease in bounce rates, suggesting users find immediate value upon entering the platform.

Response by the LLM (GPT-4):

Evaluation of proposed decision

- **Alignment with goals and premises**
    - The integration of an AI-driven personalization engine aligns with the goal to increase user engagement by tailoring content to user preferences.
    - Leveraging AI to analyze user behavior and historical data fits the premise that personalized content can enhance user engagement.

GUIDING QUESTIONS FOR STRATEGY DEVELOPMENT

Guiding questions for strategy development

- ***Understanding user value***
    - How might we ensure that the AI-driven personalization respects user privacy and maintains trust while delivering tailored content?
    - In what ways can we involve users in the personalization process to enhance their sense of control and ownership over their data?
- ***Long-term sustainability***
    - What measures can we take to continuously improve the AI's accuracy and relevance over time without compromising ethical standards?
    - How can we design the personalization engine to adapt to changing user preferences and market conditions?

Evaluation of expected early indicators

- ***Relevance to proposed decision***
    - The expected early indicators are directly related to the proposed decision and provide measurable outcomes that reflect increased user engagement.
    - Positive user feedback and a decrease in bounce rates are strong indicators of the value provided by the personalized content.

Guiding questions for measuring success

- ***KPIs and metrics***
    - What key performance indicators (KPIs) can we establish to measure the impact of personalized content on user engagement beyond click-through rates and session durations?
    - How can we quantify user satisfaction and the perceived value of AI-recommended content?
- ***Experimentation and testing***
    - What experiments can we conduct to test the effectiveness of personalized content in real time and make iterative improvements?

- *How can we validate that the AI-driven personalization engine is enhancing the user experience for different segments of our user base?*

By considering these questions, you can guide your coachee towards developing a more robust, sustainable, and user-centric product strategy that aligns with the overarching goals and premises.

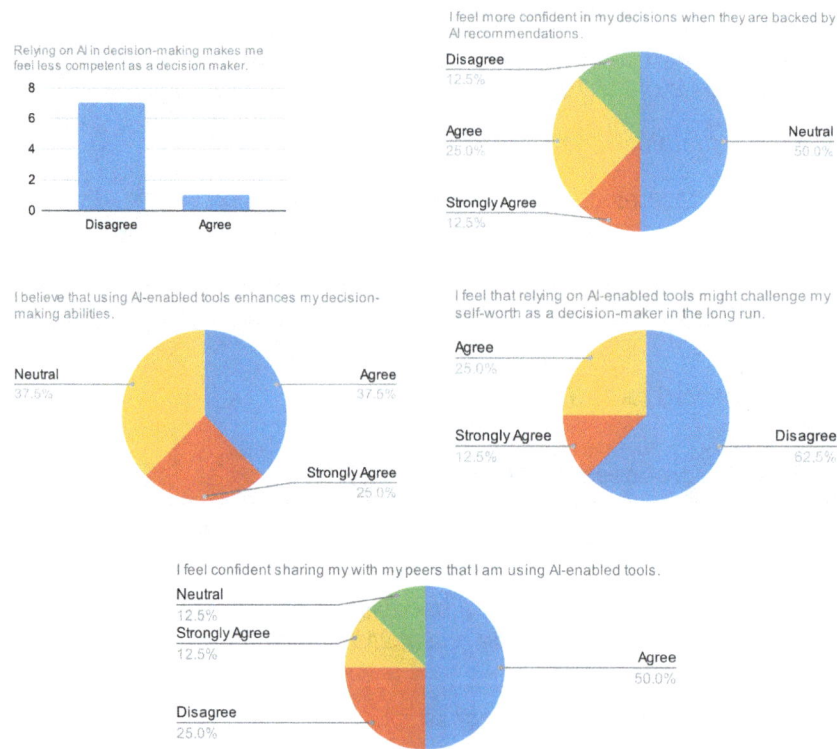

It was not possible for the participants to follow up on the response they received. After this, the participants were invited to fill out an online survey about how the use of these AI-enabled tools impacted their self-image as decision makers. The survey was constructed out of statements to which the participants were asked to respond on a Likert scale (Willits, Theodori, and Luloff 2016). Eight out of nine participants completed and submitted the survey. Notably, only three out of the eight participants agreed to having significant experience using AI-enabled tools. The responses suggest a relative lack of concern about losing competence, but rather high uncertainty about potential gains in confidence, even though there is a tendency to believe that abilities in decision making can benefit from AI-enabled tools. This is, however, again met with uncertainty about the participant's self-worth as a decision maker in the long run.

With a small number of participants, no general findings can be derived from the survey. However, there is a sense that participants do not feel they stand to lose much by using the AI-enabled tools provided to them in this set of workshops. On the positive side, participants do seem to value the capabilities added to decision making by using AI-enabled tools. From the qualitative results of the workshop preceding the survey, we can gather that the AI-enabled tool providing questions to the decision model from its artificial perspective was seen as contributing significant value specifically because the perspective was different from that of the participant. This is where future research and development can find its path: AI-enabled tools are already seen as sufficiently independent in their artificial perspective to facilitate a notion of collaboration with an additional team member. The AI-enabled tools thus have the potential to provide collaboration opportunities with artificial perspectives with a wide range of customizability, as every artificial perspective can be tuned to match the goals of the collaboration at hand. Detailed form analysis and creation of decision models, as outlined above, can potentially serve as enablers in the creation of such collaborative artificial perspectives.

9 Conclusion

We have explored various social aspects of decision making, including the construction of decision models, with a specific focus on scenario-based decision models. While this might have created the impression of very structured decision making processes, reality is always (necessarily) more complex than any model and process. This should become clear via the references to framing biases, for example. It is here that a rich field of empirical data presents itself for future research, especially in combination with new "artificial" perspectives on decision making brought about by the widespread use of LLM-based AI tools. In addition, we have argued that decision models are not only less *structured* than expected in classical literature, but also have a much more *structuring* function in processes of organizing work and the communication with stakeholders via the surplus of meaning they create and which organizations have to process.

References

[1] Baecker, D. 2019. "Negation and Imagination in Economic Calculus." SSRN Scholarly Paper. Rochester, NY. https://doi.org/10.2139/ssrn.3500713.

[2] Baecker, D. 2022. "Die Spencer-Brown-Transformation." https://doi.org/10.13140/RG.2.2.18000.58888/1.

[3] Dixson-Decleve, S., Gaffney, O., Ghosh, J., Randers, J., Rockstrom, J. and Stoknes, P. E. 2022. *Earth for All: A Survival Guide for Humanity*. Gabriola Island, British Columbia, Canada: New Society.

[4] Esposito, E. 2022. *Artificial Communication: How Algorithms Produce Social Intelligence*. Cambridge, MA; London: The MIT Press.

[5] Federal Aviation Administration. 2017. "AC120-71B: Standard Operating Procedures and Pilot Monitoring Duties for Flight Deck Crewmembers." U.S. Dept. of Transportation, Federal Aviation Administration.

[6] von Foerster, Heinz 2003. "Ethics and Second-Order Cybernetics." In *Understanding Understanding*, 287—304. New York: Springer.

[7] Foley, C. F., and Khavkin, M. 2019. "How Companies Should Prepare Their Forecasts." *Harvard Business Review*, 15 April 2019. https://hbr.org/2019/04/how-companies-should-prepare-their-forecasts.

[8] Gödel, K. 1931. "Über formal unentscheidbare Sätze der Principia Mathematica und verwandter Systeme I." *Monatshefte für Mathematik und Physik* 38 (1): 173–98. https://doi.org/10.1007/BF01700692.

[9] Graham, T. 2018. "Platforms and Hyper-Choice on the World Wide Web." *Big Data & Society* 5: (1). https://doi.org/10.1177/2053951718765878.

[10] Grote, F. 2014. *Locating Publics: Forms of Social Order in an Electronic Music Scene*. Wiesbaden: Springer.

[11] Grote, F. 2023. "Distinction Dynamics: A Form Analysis of Self-Descriptions in Agile Teams." *Soziale Systeme* 28 (1): 130–62. https://doi.org/10.1515/sosys-2023-0008.

[12] Hagen, J U., Lei, Z. and Shahal, A. 2019. "What Aircraft Crews Know About Managing High-Pressure Situations." *Harvard Business Review*, 9 December 2019. https://hbr.org/2019/12/what-aircraft-crews-know-about-managing-high-pressure-situations.

[13] Johnson, E. J. 2022. *The Elements of Choice: Why the Way We Decide Matters*. New York: Penguin.

[14] Kahneman, D. 2013. *Thinking, Fast and Slow.* 1st pbk. ed. New York: Farrar, Straus & Giroux.

[15] Korenhof, P., Giesbers, E. and Sanderse, J. 2023. "Contextualizing Realism: An Analysis of Acts of Seeing and Recording in Digital Twin Datafication." *Big Data & Society* 10 (1): 20539517231155061. https://doi.org/10.1177/20539517231155061.

[16] Luhmann, N. 1995. *Social Systems.* Stanford: Stanford Univ. Press.

[17] Luhmann, N. 2017. *Die Kontrolle von Intransparenz.* Berlin: Suhrkamp.

[18] Luhmann, N. 2018. *Organization and Decision.* Translated by Baecker, D. and Barrett, R. Cambridge: Cambridge Univ. Press.

[19] McFee, B. 2022. "Digital Sampling." In *Digital Systems Theory.* https://brianmcfee.net/dstbook-site/content/ch02-sampling/intro.html.

[20] Meadows, D. H., Meadows, D. L., Randers, J. and Behrens, W. W. 1972. *Limits to Growth. A Report for the Club of Rome's Project on the Predicament of Mankind.* New York: Universe Books.

[21] Meadows, D. H., and Randers, J. 2013. *Limits to Growth: The 30-Year Update.* White River Junction: Chelsea Green.

[22] OpenAI. 2023. "GPT-4." 14 March. https://openai.com/research/gpt-4.

[23] Pearl, J., and Mackenzie, D. 2018. *The Book of Why: The New Science of Cause and Effect.* 1st ed. New York: Basic Books.

[24] Richter, T., and Groth, T. 2023. *Wirksam führen mit Systemtheorie: Kernideen für die Praxis.* 1st ed. Heidelberg: Carl-Auer.

[25] Rogers, A. 2023. "The Top 50 Books Being Used to Train ChatGPT—and What They Say about Its 'Intelligence.'" *Business Insider.* 30 May. https://www.businessinsider.com/chatbot-training-data-chatgpt-gpt4-books-sci-fi-artificial-intelligence-2023-5.

[26] Rosenzweig, P. 2014. "The Benefits—and Limits—of Decision Models." *McKinsey Quarterly*, 1 February. https://www.mckinsey.com/capabilities/strategy-and-corporate-finance/our-insights/the-benefits-and-limits-of-decision-models#/.

[27] Schaul, K., Chen, S. Y., and Tiku, N. 2023. "Inside the Secret List of Websites That Make AI like ChatGPT Sound Smart." *Washington Post.* 19 April. https://www.washingtonpost.com/technology/interactive/2023/ai-chatbot-learning/.

[28] Simon, H. A. 1972. "Theories of Bounded Rationality." In *Decision and Organization*, edited by C. B. McGuire and Roy Radner, 161–76. Amsterdam: North-Holland.

[29] Spencer-Brown, G. 1969. *Laws of Form*. London: George Allen & Unwin.

[30] Spencer-Brown, G. 2008. *Laws of Form*. 5th English Edition. Leipzig: Bohmeier.

[31] Sterman, J. D. 2000. *Business Dynamics: Systems Thinking and Modeling for a Complex World*. Boston, MA; London: McGraw-Hill.

[32] Tetlock, P., and Gardner, D. 2015. *Superforecasting: The Art and Science of Prediction*. 1st ed. London: Cornerstone Digital.

[33] Weick, K. E. 2000. *Making Sense of the Organization*. Malden, MA; Oxford: Wiley-Blackwell.

[34] Weick, K. E. 2010. "Reflections on Enacted Sensemaking in the Bhopal Disaster." *Journal of Management Studies* 47 (3): 537–50. https://doi.org/10.1111/j.1467-6486.2010.00900.x.

[35] Weisstein, E. W. 2003. "Logistic Equation." Text. Wolfram. https://mathworld.wolfram.com/.

[36] Willits, F. E., Theodori, G. and Luloff, A. 2016. "Another Look at Likert Scales." *Journal of Rural Social Sciences* 31(3). https://egrove.olemiss.edu/jrss/vol31/iss3/6.

Laws of Form and Husserl's "Strange World of the Purely Logical"

Claire Ortiz Hill

claireortizhill@gmail.com

Abstract

Connections are made between George Spencer-Brown's *Laws of Form* and Edmund Husserl's theories about what he called the "strange world of the purely logical." Specifically, I talk about: Husserl's background as a mathematician and his use of symbolic notation. Then, I compare his and Spencer-Brown's views on: the relationship between mathematics and logic, mentioning Boole; the fundamental particles from which numbers can be made; the structure of knowledge of the universe, especially Husserl's theory of manifolds; imaginary entities; and Russell's theory of types. I suggest that Spencer-Brown's and Russell's brief exchange on propositional functions and Husserl's theories about them is important and calls for independent treatment.

Interesting connections are to be drawn between Edmund Husserl's theories about what he called the "strange world of the purely logical" and George Spencer-Brown's theories about the same in *Laws of Form*. Indeed, Husserl was not really the lopsided philosopher commonly perceived by followers and foes alike to have spent most of his life celebrating only about subjectivity. However, the evidence of his years-long work to lay bare and impart knowledge about the true and ultimate structure of objective reality became buried in the excitement generated by his extensive exploration of the realm of transcendental subjectivity, with which he was so smitten that he did not pursue his theories about pure logic as far as he could have. Indeed, in 1917, he wrote to Hermann Weyl of how, despite all the work he had devoted to formal logic, he had not followed it completely to the end, because it had had to be more important to him to develop his ideas about transcendental phenomenology. In 1930, he wrote to Georg Misch of how he had lost all interest in formal logic and all real ontology in the face of a systematic grounding of a theory of transcendental subjectivity.

Zealous followers followed suit. So, Husserl's pioneering work in formal logic and ontology has not received the attention it deserves, even, or especially, from his most ardent followers. However, he left intact his austere scheme to come to clarity with respect to the central traits of reality and, since the 1980s, the Husserl Archives has been publishing his lecture courses on logic, theory of knowledge and science which provide material needed to retrieve the map of the world of the purely logical he drew (Husserl 2001b; 2001c; 2003a; 2008; and especially 2019). So, lovers of *Laws of Form* can now experiment with his scarcely explored theories about what Spencer-Brown called "a reality independent of how the universe actually appears." If, as Spencer-Brown suggested, the initial exploration of the basic forms underlying knowledge is usually undertaken in the company of an experienced guide, then Husserl can be such a guide and his ideas can be what Spencer-Brown called "a key to a world beyond the compass of ordinary description" (Spencer-Brown 2011, xxii).

Against this background, I want to talk about some connections I find between Husserl's and Spencer-Brown's theories. Given the limitations imposed, I must aim for pithiness. I hope that not too much gets lost in the wash.

First, it is essential to know that Husserl was a mathematician by training, who for decades kept close company with the most outstanding and pioneering mathematicians of his time, namely Karl Weierstrass, Georg Cantor and David Hilbert and his circle (Hill 2002a). He was as knowledgeable, if not more so, about the latest developments in mathematics and logic as were the makers and champions of modern logic and mathematics, say Gottlob Frege, Bertrand Russell, and Willard Quine, who were embraced by the Anglo-American philosophical establishment, which controlled those fields, and philosophy in general, in English-speaking countries during the 20th century.

Second, Spencer-Brown stated that his "conscious intention" in writing *Laws of Form* had been "the elucidation of an indicative calculus" (Spencer-Brown 2011, xix). Husserl obviously could not have used that notation, but it is not well known that he did use symbolic notation (See Hill 2016). It figures in his lectures on the pure mathesis in his logic courses of 1902/03 and 1896 (e.g. Husserl 2001b, 254–64; 2001c, 239–50)—which contains a 23-page section on George Boole—and it looks like that of C. S. Peirce and Ernst Schröder, who did much to introduce Boole's work on the continent. It has been widely used and is familiar and intelligible nowadays.

Third, for Spencer-Brown a "principal intention" of *Laws of Form* was "to separate what are known as algebras of logic from the subject of logic, and to re-align them with mathematics" (Spencer-Brown 2011, xiv). Accounts of the properties of Boolean algebras, he bemoaned, had revealed "nothing of any mathematical interest about their arithmetics." And "nobody" had "made any sustained attempt to elucidate and to study the primary, non-numerical arithmetic" of Boolean algebra, so that when he himself had begun to see the need for this he had found himself "upon what was, mathematically speaking, untrodden ground" (ibid., xiv).

Yet, Spencer-Brown stressed, mathematics is a "powerful" way of "revealing our internal knowledge of the structure of the world, and only by the way associated with our common ability to reason and compute" (ibid., xv). He saw the relation of logic to mathematics as being "that of an applied science to its pure ground..." (ibid., 83). The "subject matter of logic," he wrote, "however symbolically treated is not, in as far as it confines itself to the ground of logic, a mathematical study. It becomes so only when we are able to perceive its ground as a part of a more general form, in a process without end. Its mathematical treatment is a treatment of the form in which our way of talking about our ordinary living experience can be seen to be cradled." It was the laws of that form, rather than those of logic, that he wanted to record (ibid., xx).

For Husserl, the formalization of large tracts of mathematics in the 19th century had brought the parallels between its structures and those of logic to the fore, thus raising profound new questions about deep underlying connections between the two fields. This had unmasked close relationships between the propositions of logic and number statements and enabled logicians to develop a genuine logical calculus for calculating with propositions in the way mathematicians do with numbers, quantities and the like. Mathematizing logicians had done well to recognize the essential likeness between the formal mathematical disciplines and the disciplines of logical validity and to have raised them to a higher level of technical perfection by applying the same, completely appropriate, algebraic methods to them, thus expanding the scope of the exact mathematical disciplines (Husserl 1969, Chpt. 2).

He considered traditional syllogistic logic to be a piece of pure mathematics, namely, the pure mathematics of propositions and of predicates of possible subjects in general, where, he wrote, formalization "leads to a theory-form that can be understood as a special case of the formal genus-type 'arithmetic.' All the well-known algebraic propositions $ab = ba$, the laws of association, distribution, hold," he noted, "and the brilliant Boole saw that two closed domains of ordinary syllogistic logic can be dealt with as if it were an arithmetic" (Husserl 2019, §58). So, by adopting the deductive theories of traditional syllogistic logic and gradually developing a mathesis of propositional, conceptual and relational meanings, 19th century mathematicians had but laid hold of a field that was their own.

Husserl taught that there was nothing extraordinary about the idea of calculating with concepts and propositions and *a priori* no reason why calculation should be limited to the arithmetical field because the formal discipline of propositions in general and of concepts in general was a mathematical discipline of the same nature and used the same methods as familiar mathematical disciplines like arithmetic, which was the most marvelous tool devised for purposes of deduction, the science in which the deductive relations were analyzed most carefully. The fact that one could generalize and produce variations of formal arithmetic that led outside the quantitative domain without essentially altering formal arithmetic's theoretical nature and calculational methods had shown him that there was more to the mathematical or formal sciences, or the mathematical method of calculation, than could be captured in purely quantitative analyses. For him, the essence of the mathematical did not lie in being quantitatively determinable but in establishing a purely apodictic foundation of the truths of a field from apodictic principles, a matter of a rigorously scientific, *a priori* theory building from the bottom up and deriving the manifold of possible inferences from axiomatic foundations *a priori* in a rigorously deductive way requiring one and the same method everywhere: the algebraic method. For him, the essential thing in mathematics was not the objects, but its method which naturally flows into a purely symbolic technique.[1]

Fourth, for Spencer-Brown, his own "basic contribution to mathematics" had been "to discover the fundamental particles from which numbers and other, simpler, elements of mathematical systems can be made" (Spencer-Brown 2011, 118). For

---

[1] (See, e.g Husserl 2019, §§50, 58; 2008, §§15, 19a, 442–44; 2001b, 241–42, 271–73; 2001c, 19, 34; Hill 2010a).

Husserl, the concept of number was a paradigm of a purely logical concept, namely, a concept which is not limited to a special field of objects, but relates to objects in general in the most universal ways, which not only can and does figure in all the sciences, but is common and necessary to all sciences, because it belongs to what belongs to the ideal essence of science in general. No science, he pointed out, is conceivable in which the number concepts cannot find an application. So, all purely mathematical concepts like unit, multiplicity, cardinal number, order, ordinal number, and manifold are purely logical because they clearly relate in the most universal way to numbers in general and are only made possible out of the most universal concept of object.

Since each and every thing can be counted as one, to conceive the concept of number, or any arbitrarily defined number, Husserl explained, we only need the concept of something in general. One is something in general. Each and every thing can be counted as one and out of the units all cardinal numbers built. Cardinal number is a specific differentiation of the concept of multiplicity which is the most universal logical concept combining objects in general. The first number in the number series is 2 *As*. From 2 *As*, we use definitions to form the new number 2 *As* and 1 *A* which we designate as 3 *As*, etc., then we obtain a series of the natural numbers, infinite in one direction.

To questions as to how arithmetic came about and the foundations of arithmetic provided, Husserl answered that people analyzed the arithmetic propositions as they were first entertained by people. They found that certain relations were grounded in the concept of number. For instance, any two numbers are either equal or one is larger or smaller than the other. They found that certain combinations were grounded in the concept of number: addition, multiplication, subtraction, division, etc. Given with the elementary combinations were certain simple, directly intelligible laws that careful analysis traced back to a certain minimal number of laws no longer reducible to one another, which are *a priori* since they lie in the simple meaning of the concepts founding them. They are propositions about relations of ideas obtained by analysis of the universal concepts by digging more deeply into their meaning. The unending profusion of theories that arithmetic develops is fixed, enfolded in the axioms, each of which is a proposition that systematically unfolds from some side the meaning of cardinal number, itself the answer to the question "How many?" or unfolds some of the ideas inseparably connected with it, following simple procedures. The field branches out into more and more theories and partial disciplines, new problems surface and are solved using the most rigorous methods (see e.g. Husserl 2008, §13c; 2001c, 31–43, 241–46; Hill 2010a).

Fifth, Spencer-Brown variously alluded to "investigations of the inner structure of our knowledge of the universe, as expressed in the mathematical sciences" (Spencer-Brown 2011, xviii), "our internal knowledge of the structure of the world" (ibid., xv), "mathematical form as an archetypal structure" (ibid., xvii), "the structure beyond ordinary experience in which all creation hangs together" (ibid., xxii).

For Husserl's part, he depicted the structure of the world of pure logic[2] in a way which he considered to be a radical clarification of the relationship between formal mathematics and formal logic. He detected a natural order in formal logic consisting of three levels, on the first of which he placed the traditional Aristotelian logic of subject and predicate propositions and states of affairs. There, numbers, for example, do not occur as independent objects about which something is predicated but as form, thereby dependent, as when we say: '3 houses.' Sets as objects do not occur there either because in set theory, judgments are not made directly about elements but about sets, whole totalities of elements. If we make such forms independent, new higher-order objects emerge. For him, traditional Aristotelian logic was but a small area of pure logic which had to be distinguished and segregated from the formal ontology of the broader sphere of pure logic, which included the mathematical disciplines and was immense in comparison.

In the disciplines of the two higher levels, it is a matter of investigating what is valid for higher-order object formations determined in purely formal terms, grounded in the essence of logical forms and dealing with objects in indeterminate, general ways. Husserl described the second level as an expanded, completely developed analytics in which one proceeds in a purely formal manner. One calculates, reasons deductively, with concepts and propositions. Signs and rules of calculation suffice because each procedure is purely logical. One manipulates signs, which acquire their meaning in the game through the rules of the game. One may proceed mechanically in this way and the result will prove accurate and justified. Numbers function entirely differently on this level, where statements about numbers in which numbers are the objects are found. As an example of such arithmetical propositions, he gave: "Any number can be added to any number." Here, he located the basic concepts of mathematics, the theory of cardinal numbers, of ordinals, set theory, mathematical physics, formal pure logic, pure geometry, geometry as *a priori* theory of space, the axioms of geometry as a theory of the essences of shapes, of spatial objects, but also

---

[2]I especially aimed to depict Husserl's depiction of the structure of the world of pure logic in Hill 2015b, 62–83. See also Hill 2018, 34–64. Both papers are anthologized in Hill 2024.

the pure theory of meaning and being, *a priori* real ontology of any kind, ontology of nature, ontology of minds, natural scientific ontology, sciences of value, pure ethics, the logic of morality, the ontology of ethical personalities, axiology, pure esthetics, the logic of the ideal state or ideal world government as a system of cooperating ideal nation states, the ideal of a valuable existence, essence-propositions about objects insofar as they are objective truths and as truths have their place in a truth-system in general, etc.

On the third and highest level of formal logic, Husserl placed his science of theory forms, his theory of manifolds or *Mannigfaltigkeitslehre*.[3] It was to be a new method constituting a new kind of mathematics, the most universal of all, a technique for engaging in pure *a priori* analyses through an austere scheme of axiomatization, a matter of theorizing about possible fields of knowledge conceived of in a general, undetermined way and simply determined by the fact that the objects stand in certain relations, themselves subject to certain fundamental laws of such and such determined form, a science of deductive systems in general, a field of free, creative investigation made possible once form was emancipated from content. One is free to reason completely on the level of pure forms where the systems can vary in different ways. One finds ways of constructing an infinite number of forms of possible disciplines.

His manifolds were to be pure forms of possible theories which, like molds, are totally undetermined as to their content and not bound to any concrete interpretation, but to which thought must conform in order to be thought and known in a theoretical manner. Only a form is defined. It exists insofar as it is correctly defined, insofar as the axiom forms are ordered in such a way as to contain no formal contradictions, no violation of analytic principles. For example, one speaks of numbers in the formal sense, but one does not mean cardinal numbers, quantitative numbers, or anything of the kind but anything for which formal axioms of the arithmetical prototype hold. If we drop the cardinal number meaning of the letters in the ordinary theory of cardinal numbers and substitute the thought of objects in general for which axioms of arithmetical forms are to hold, we no longer have arithmetic but a purely logical class prototype of theory forms to which, besides innumerably many possible domains, the domain of cardinal numbers is also subject. For cardinal numbers, $ab = ba$ holds. In constructing a manifold, though, one may just as well stipulate that $ab \neq ba$.

---

[3]Husserl mainly discussed his theory of manifolds in Husserl 1970, §§69–70; 1901a; 2023b, 409–57; 2008, §19; 1962, §§71–72; 2019, §§54–59; 1969, §33.

He suggested this meaning for the theory of non-Euclidean manifolds: "Let there be a domain in which the objects are subject to certain forms of relation and connection, for which axioms of such and such a form are valid, then for a domain formally constituted in this way, a mathematics of such and such a form would be valid, there would then result propositions of such and such a form, proofs, theories of such and such a form. *There is no domain. There are no actually given concepts, connections, relations and axioms.* One simply says, *if* one had a domain, and *if* axioms of such and such a form obtained for it" (Husserl 2008, §19d).

Husserl believed that it was up to mathematicians, as the only competent engineers of deductive structures, to construct such theories and theoretical disciplines, while it was up to philosophers to engage in complementary reflections on the essence and meaning of the basic concepts and laws. He believed that the idea of a theory of forms of meanings as a discipline prior to the disciplines of logical validity was still completely beyond the ken of mathematizing logicians (ibid.).

Sixth, Spencer-Brown stressed that by failing to make use of the crucially important fact that a calculus and its interpretation are distinct, one cuts oneself off from readily available forms of simplification, one of them, frequent in mathematics, being the underlying use of a construction that is devoid of interpretation in the particular application, but can be used to shorten the way to an answer in it (Spencer-Brown 2011, 91). As an example, he gave the $\sqrt{-1}$. He said that "perhaps the most significant thing" about his calculus "from the mathematical angle" was that it enabled one to use complex values in the algebra of logic, which are "the analogs, in ordinary algebra, to complex numbers" where they "are accepted as a matter of course," the "more advanced techniques" being "impossible without them." Yet in Boolean algebra, and thus "in all our reasoning processes, we disallow them." He had seen how their Boolean counterparts worked perfectly well in practical engineering but, like the first mathematicians to use 'square roots of negative numbers,' had felt guilty about using them because he "could see no plausible way of giving them respectable academic meaning." However, he was "quite sure there was a perfectly good theory that would support them," if only he could think of it (ibid., x). So, he extended the concept of imaginaries to Boolean algebras, so that a valid argument might contain four classes of argument: true, false, meaningless and imaginary, something he saw as having profound implications for mathematics, logic, philosophy and even physics (ibid., xi). That imaginary values could be used to reason towards a real and certain answer, combined with the fact that they *were not being* used so in mathematical reasoning and with the fact that certain equations plainly could not be solved without using them, he reasoned, "meant that *there **must** be mathematical* *statements (whose truth or untruth is in fact perfectly decidable) that cannot be decided by the methods of reasoning to which we have hitherto restricted ourselves."* If "we confine our reasoning to an interpretation of Boolean equations of the first degree only, we should expect to find theorems that will always defy decision, and the fact that we do seem to find such theorems in common arithmetic may serve, here, as a practical confirmation of this obvious prediction" (ibid., 81).

For his part, Husserl maintained that his chief purpose in developing his theory of manifolds had been to find a theoretical solution to the thitherto unclarified problem of imaginaries, to how in the realm of numbers, impossible, non-existent, meaningless concepts could be dealt with as real ones. In the early 1890s, he wrote to Carl Stumpf of how, in trying to understand how operating with contradictory concepts could lead to correct theorems, he had found that for numbers like $\sqrt{2}$ and $\sqrt{-1}$, it was not a matter of the possibility or impossibility of concepts. Through the calculation itself and its rules as defined for such numbers, the impossible fell away and a genuine equation remained. One could calculate again using the same signs, but referring to valid concepts, and the result was again correct. The calculation remained correct if it followed the rules even if one wrongly imagined that what was contradictory existed or held the most absurd theories about the content of the corresponding concepts of number. So, this must be a result of the signs and their rules (Husserl 1994, 12–19).

He saw his theory of manifolds as the key to the only possible solution to the problem of imaginaries. Understanding the nature of theory forms had shown him how reference to impossible objects could be justified. It was formal constraints banning meaningless expressions, meaningless imaginary concepts, reference to non-existent and impossible objects that was restricting theoretical, deductive work, but resorting to the infinity of pure forms and transformations of forms freed one from such conditions and explained why having used imaginaries, what is meaningless, must lead, not to meaningless, but to true results. There are no negative numbers in the arithmetic of cardinal numbers because the meaning of the axioms is so restrictive as to make subtracting 4 from 3 nonsense. Irrational numbers, $\sqrt{-1}$ are meaningless there, and so on. We cannot arbitrarily expand the concept of cardinal number, but we can abandon it and define a new, pure formal concept of positive whole number with the formal system of definitions and operations valid for cardinal numbers. And, this formal concept of positive numbers can be expanded by new definitions while remaining free of contradiction. One can operate freely within a manifold with imaginary concepts and be sure that what one deduces is correct when the axiom system completely and unequivocally determines all the configurations possible in a domain through a purely analytical procedure. It is the completeness of the axiom system that gives one the right to operate freely. A domain is complete when each grammatically constructed proposition solely using the language of the domain is, from the outset, determined to be true or false in virtue of the axioms.[4]

Seventh, "To make their mathematical logic conform with the logic of Aristotle," Spencer-Brown thought, Russell and Whitehead introduced the theory of types, "an arbitrary principle... that effectively forbade all arguments that would have to be represented by equations of degree higher than unity." They mistakenly introduced it, he explained, "expressly to disallow complex values," so that "in this field, the most advanced techniques, though not impossible, simply did not yet exist," and one was still constrained in one's reasoning processes to proceed as in Aristotle's time. However, all that had to be done, he believed, was to show that the self-referential paradoxes discarded with the theory were no worse than similar self-referential paradoxes considered acceptable in the ordinary theory of equations and thus resolvable by introducing imaginaries. It was possible to solve equations of higher degree that the theory of types had excluded from ordinary logic and that had been undertaken. He said that when, in 1967, he showed Russell the proof that his theory was unnecessary, he abandoned his belief in it and admitted that it was "the most arbitrary thing" that he and Whitehead had ever had to do and was glad to see the matter resolved. However, though Russell and Spencer-Brown agreed that a mathematical proof with a logical argument requiring an algebraic equation of higher degree was possible, neither of them could then imagine what it might look like (Spencer-Brown 2011, x-xi, xviii, 198).

In any case, Russell had recognized decades earlier that the theory of types could not be "the key to the whole mystery" of the onerous contradiction he had worked so hard to evade. He saw that deeper problems caused it to break out afresh and "further subtleties... needed to solve them" (Russell 1919, 135; 1956, 333). After all, the theory was but an *ad hoc* attempt to evade the contradictions derivable in Frege's system by restoring the formal structure that he programmed his system to break. Frege had placed the need to lay hold of self-subsistent, independent, objects at the heart of his system in a way that forced him to devise a law to let him pass from a concept to its extension, something which he believed was "forbidden by the basic difference between first and second level relations." Yet, he temporarily convinced himself that he might assume "an unprovable law" to legitimate the illicit transformation (See, e.g. Frege 1979, 182, 269–70; 1980, 54–55).

---

[4]See references in Footnote 3, as well as Hill 2002b.

Russell said that the contradiction had taught him that classes could not be independent entities in the same sense in which things are things, that if a word or phrase that is devoid of meaning is wrongly assumed to have an independent meaning, false abstractions, pseudo-objects, paradoxes and contradictions were apt to result. Along with his theory of definite descriptions, no-classes theory, axiom of reducibility, the theory of types was one of his various attempts to get rid of the contradiction-generating imaginary objects that Frege's system generates. However, he believed that "without a single object to represent an extension mathematics crumbles," (Russell 1903, §489; Hill 2004, 207–32)—as if mathematics would crumble because some thinkers had devised some bad theories.

In comparison, for Husserl, as seen, sets and classes were not independent entities occurring in traditional Aristotelian logic, but higher-order object formations figuring on the second level of formal logic, because in them, judgments are not made directly about elements, but about sets, whole totalities of elements. In fact, Frege, Russell, and Husserl all concluded that the essential differences between dependent and independent meanings were of the highest importance, inviolable and "founded deep in the nature of things," so that antinomies, contradictions, paradoxes, fallacies, nonsense, confusion, absurdity, mysteries inevitably result were they not respected. Specifically, insidious problems with pseudo-objects, inference, existential generalization, type ambiguities, substitutivity of identity, semantical paradoxes, namely much of what analytic philosophers have been battling since Frege's time, inevitably creep into reasoning (see, e.g. Hill 2010b, 313–32; 2003). Frege's logic, in fact, led to what he himself called a thicket of contradictions, what Russell called a bewildering maze, to Quine's fragmented world of rabbit parts, stages, and fusions; river stages; kinship; and person stages, where the ontologies of physical and mathematical objects are but myths relative to an epistemological view (Quine 1947; 1956; 1960), not to mention the deep confusion that, for a while, during the 1980s, analytic philosophers admitted to experiencing.

As an eighth connection, I might propose reflecting on Spencer-Brown's and Russell's brief exchange regarding propositional functions (Spencer-Brown 2011, 114) in connection with Husserl's theories about the same (see Hill 2016), a highly interesting subject, which requires more in-depth treatment than is possible here.

References

[1] Frege, G. 1979. *Posthumous Writings*, Oxford: Blackwell.

[2] Frege, G. 1980. *Philosophical and Mathematical Correspondence*, abridged by McGuinness, B., Gabriel, G. et al. (eds.), Oxford: Blackwell.

[3] Hill, C. O. 2002a. "On Husserl's Mathematical Apprenticeship and Philosophy of Mathematics," in *Phenomenology World Wide*, Tymieniecka, A-T (ed.), Dordrecht: Kluwer, 76–92, anthologized in Hill & da Silva 2013.

[4] Hill, C. O. 2002b. "Tackling Three of Frege's Problems: Edmund Husserl on Sets and Manifolds," *Axiomathes, an International Journal in Ontology and Cognitive Systems*, 13(1): 79–104, anthologized in Hill & da Silva 2013.

[5] Hill, C. O. 2003. "Incomplete Symbols, Dependent Meanings, and Paradox," in *Husserl's Logical Investigations*, Dahlstrom, D. O. (ed.), Dordrecht: Kluwer, 69–93, anthologized in Hill & da Silva 2013.

[6] Hill, C. O. 2004. "Reference and Paradox," *Synthese*, 138(2): 207–32, anthologized in Hill & da Silva 2013.

[7] Hill, C. O. 2010a. "Husserl on Axiomatization and Arithmetic," in *Phenomenology and Mathematics*, Hartimo, M. (ed.), 47–71, Dordrecht: Springer, anthologized in Hill and da Silva 2013.

[8] Hill, C. O. 2010b. "On Fundamental Differences Between Dependent and Independent Meanings," *Axiomathes, An International Journal in Ontology and Cognitive Systems* 20: 2–3, online since May 29, 2010, 313–32 (DOI 10.1007/s10516-010-9104-1), anthologized in Hill & da Silva 2013.

[9] Hill, C. O. 2015a. "Husserl's Way Out of Frege's Jungle," in *Objects and Pseudo-Objects Ontological Deserts and Jungles from Brentano to Carnap*, Leclercq, B., Richard, S. and Seron, D. (eds.), Berlin: de Gruyter, 183–96, anthologized in Hill 2024.

[10] Hill, C. O. 2015b. "The Strange Worlds of Actual Consciousness and the Purely Logical," *New Yearbook for Phenomenology and Phenomenological Philosophy* 13: 62–83, anthologized in Hill 2024.

[11] Hill, C. O. 2016. "Husserl and Frege on Functions," in Rosado Haddock 2016, 89–117.

[12] Hill, C. O. 2018. "Limning the True and Ultimate Structure of Reality," in *Mereologies, Ontologies, and Facets: The Categorial Structure of Reality*, Hackett, P. (ed.) Lanham MD: Lexington Books, 34–64, anthologized in Hill 2024.

[13] Hill, C. O. 2024. *Experience and the Ultimate Structure of Reality, Husserl's Pursuit of Truth*, London: College Publications.

[14] Hill, C. O. and Rosado Haddock, G. 2000. *Husserl or Frege? Meaning, Objectivity, and Mathematics*, La Salle IL: Open Court.

[15] Hill, C. O and da Silva, J. J. 2013. *The Road Not Taken, On Husserl's Philosophy of Logic and Mathematics*, London: College Publications.

[16] Husserl, E. 1901a. "Double Lecture: On the Transition through the Impossible (Imaginary) and the Completeness of an Axiom System," in his *Philosophy of Arithmetic*, 409–57.

[17] Husserl, E. 1901b. "Essay IV, The Domain of an Axiom System/Axiom System—Operation System," in Husserl 2003, 475–92.

[18] Husserl, E. 1962. *Ideas, General Introduction to Pure Phenomenology*, New York: Colliers, first published, 1913.

[19] Husserl, E. 1969. *Formal and Transcendental Logic*, The Hague: Martinus Nijhoff, first published, 1929.

[20] Husserl, E. 1970. *Logical Investigations*, London: Routledge & Kegan Paul, first published, 1900/01.

[21] Husserl, E. 1973 *Experience and Judgment*, London: Routledge and Kegan Paul, first published, 1939.

[22] Husserl, E. 1975. *Introduction to the Logical Investigations, A Draft of a Preface to the Logical Investigations*, Fink, E. (ed.), The Hague: M. Nijhoff, first published, 1913.

[23] Husserl, E. 1994. *Early Writings in the Philosophy of Logic and Mathematics*, Willard, D. (tr.), Dordrecht: Kluwer.

[24] Husserl, E. 2019. *Logic and General Theory of Science, Lectures 1917/18, with supplementary texts from the first version of 1910/11*, Cham, Switzerland: Springer Verlag, my translation of *Logik und allgemeine Wissenschaftstheorie, Vorlesungen 1917/18, mit ergänzenden Texten aus der ersten Fassung 1910/11*, Dordrecht: Kluwer, 1996.

[25] Husserl, E. 2001a. *Allgemeine Erkennthistheorie, Vorlesung 1902/03*, Schuhmann, E. (ed.), Dordrecht: Kluwer.

[26] Husserl, E. 2001b. *Logik, Vorlesung 1896*, Schuhmann, E. (ed.), Dordrecht: Kluwer.

[27] Husserl, E. 2001c. *Logik, Vorlesung 1902/03*, Schuhmann, E. (ed.), Dordrecht: Kluwer.

[28] Husserl, E. 2003a. *Alte und neue Logik, Vorlesung 1908/09*, Schuhmann, E. (ed.), Dordrecht: Kluwer.

[29] Husserl, E. 2003b. *Philosophy of Arithmetic, Psychological and Logical Investigations with Supplementary Texts from 1887-1901*, Willard, D. (tr.), Dordrecht: Kluwer.

[30] Husserl, E. 2008. *Introduction to Logic and Theory of Knowledge, Lectures 1906/07*, Dordrecht: Springer, my translation of *Einleitung in die Logik und Erkenntnistheorie, Vorlesungen 1906/07*, Dordrecht: Martinus Nijhoff, 1984.

[31] Husserl, E. Ms A I 35. Untitled, undated manuscript on set theory available at the Husserl Archives in Cologne, Leuven, and Paris, partially published in German by Ierna, C. and Lohmar, D. as "Husserl's Manuscript A I 35," in Rosado Haddock 2016, 289–319.

[32] Quine, W. 1947. "The Problem of Interpreting Modal Logic," *Journal of Symbolic Logic*, 12(2): 43–47.

[33] Quine, W. 1956. "Quantifiers and Propositional Attitudes," *Journal of Philosophy*, 53: 177–87.

[34] Quine, W. 1960. *Word and Object*, Cambridge MA: MIT Press.

[35] Rosado Haddock, G. (ed.). 2016. *Husserl and Analytic Philosophy*, Berlin: de Gruyter.

[36] Russell, B. 1903. *Principles of Mathematics*, London: Norton.

[37] Russell, B. 1919. *Introduction to Mathematical Philosophy*, London: George Allen & Unwin.

[38] Russell, B. 1956. *Logic and Knowledge, Essays 1901-1950*, London: George Allen & Unwin.

[39] Spencer-Brown, G. 2011. *Laws of Form*, Leipzig: Bohmeier Verlag, first published, 1969 by George Allen & Unwin.

Universal Ontology and the First Distinction: Spencer-Brown, Husserl, and Conrad-Martius

Randolph Dible

randolphdible@gmail.com

"Just supposing (as we might) that the ultimate reality, the basic ground, as it were, that renders everything exactly as it is, is something so incredibly sensitive—like a sort of infinitely fast film—that the minutest outside probe, of any kind, obscures it so that we cannot see it. If this were so, either we should never know it at all, or we should have to find a totally different way to approach it."

Only Two Can Play This Game, 30–31

"I realized that the only 'thing' (i.e. nonthing) that would be sensitive enough to be influenced by a stimulus so weak that it didn't exist, was nothing itself. That is, nothing is the only 'thing' that is so unstable that it can 'go off' of its own accord, the only 'thing' sensitive enough to be changed by nothing."

A Lion's Teeth, 148

In a chapter of *Laws of Form: A Fiftieth Anniversary* (2023), "First Philosophy and the First Distinction: Ontology and Phenomenology of *Laws of Form*," I contributed some ways of generally tying the paradigm of Spencer-Brown from its first principle, the first distinction, to first philosophy, as broadly understood from Ancient Greek philosophy to contemporary phenomenology and cybernetic philosophy. In this next development, I aim to connect this work on first philosophy to a work of ultimate philosophy. This will take the form of articulating a new universal ontology of reality and ultimate reality, based on the formal-ontological paradigm of *Laws of Form*. This project is based on the fact that the ultimate ground of the formal ontology of *Laws of Form* is the same ground as the ultimate term of the metaphysical material ontology, called universal ontology, that we will explore here.

Universal ontology is the ultimate intention of first philosophy. It brings in its wake the speculative metaphysical project of grounding all things in a common foundation, in common terms, somewhat paradoxically including in the category of all things the kinds of things that are not things at all. Such universality stands on a hypothesis of true reality that entails a true metaphysical system and its connection to a hypostatic series of grounds as real realities: antepenultimate reality, penultimate reality, and finally, the ultimate reality from which it all originates. Such a project should be understood in conventional as well as novel terms, as it seeks to approximate the cognition of a vision, method, and doctrine appropriate to the ultimate truth of being: it should present a theory of being (an ontology) including, along with all beings, beings that are beyond being. The theory itself should be a theory of forms (a phenomenology) that is also, as Husserl envisions it, a theory of possible theory-forms. The universal ontologies of the phenomenologists Edmund Husserl and Hedwig Conrad-Martius are here synthesized with George Spencer-Brown's paradigm to contribute such a system.

The novelty of Spencer-Brown's breakthrough in mathematics is captured in its central notion of the first distinction. The idea of the first distinction is the summoning of the most primordial act of consciousness and of being in the act of crossing a distinction that is both present and super-present, both simple and ultra-simple, through the framework of the idea of indication or reference, including all sorts of implicit and explicit self-reference. Spencer-Brown's Calculus of Indications is a way of seeing all the modes of the indication of the act of crossing the first distinction as all the multifarious varieties of possible indicational forms that appear in the foreground as well as forms that do not appear because they lie in the background. By positing the idea of a distinction that reaches all the way to ultimate reality at its limit station of penultimate reality and by positing the idea of indication in a calculus that expresses the unity of mathematical disciplines, Spencer-Brown made the breakthrough to a universal formal ontology. As luck would have it, the idea of the first distinction is intuitive enough to be understood as a ubiquitous key to the portal of reality standing in the center of the category of ultimate reality.

1 George Spencer-Brown's All-Encompassing Formal Ontological Paradigm

Laws of Form gives expression to a radical universality, and this universality is what I will refer to as Spencer-Brown's paradigm. Distinction, or difference, is ubiquitous and universal. But what is distinction? Clearly, in very general terms, many things are distinct as much as many things are the same. George Spencer-Brown's idea of the first distinction, and the idea of indication (or reference) connected with it, and of the form of distinction, give us a powerful economy of principles. Or more precisely, it is an economy of the form that the first principle takes such that it also gives us the inferential engine of its motivation, sustenance, and goal: the Calculus of Indications. This constellation of first ideas can be understood in formal-ontological language as the Calculus of Indications but also the ideas of indication, and of distinction, that the Calculus of Indications depends upon. The first distinction and the Calculus of Indications are interdependent. This interdependence is the first circle of being, a self-referential self-transcendence that is the seed of a series of concentric spheres that lie immanent to the activity of self-constitution. What I want to contribute to this formal ontological thesis is the universal-ontological expression of Spencer-Brown's paradigm as a kind of system appearing in our crystal ball, which we will soon come to recognize as the infinite sphere.

Whatever can be thought can be distinguished, given the formal-transcendental nature of cognition, but one also hypothesizes that objectivity can be distinguished independently of cognitive activity. That is to say that we are free to literally draw a distinction in the materiality of sand or paper or any background against which a form appears to stand in relief, and we are free to assume that this activity of distinguishing is a kind of analogy for the constitution of objective reality. But beneath all activity of distinguishing, there exists also the distinct possibility a convergence upon the limit of such freedom, reaching beyond material reality and beyond mere analogy. Spencer-Brown's idea of the first distinction reaches this radical limit through the following thought experiment. Like the phenomenological *epoché*, this experiment reduces the multiplicity of distinctions that make up the manifold of the world to the one omnipresent distinction—the first distinction—the essence of all form. This transcendental distinction stands in radical isolation from any other possible distinction, but it is necessarily accompanied by an even deeper reality beyond it. This state on the far side, beyond even the first distinction, is what Spencer-Brown calls "the unmarked state." As ideas that are hypothesized, or in Spencer-Brown's words "take[n] as given" (2011, 1), at the beginning of *Laws of Form*, these ideas have an ambiguous sense that soon becomes more definite and then continues to develop a sophisticated formal-ontological sense as this sense functionally expands into an indicational space within which the forms of Spencer-Brown's calculus develop. This calculus grows into its own indicational space out of its own ground of the act of crossing the first distinction. As self-transcendence, the hypothesis of the first distinction is the dynamism at the root of its own calculus-building activity, which in turn is the science immanent to natural dynamics. In brief, the two meanings of the word calculus, the mathematical sense and the physical sense, converge.

Laws of Form is a work of pure mathematics, but it is not only purely a work of pure mathematics. It opens with a quote from the *Tao Te Ching*, it references Proclus, and it contains hints of a metaphysical sense of its central element throughout the main text, with further reflections in the notes and appendices. In other works, such as *Only Two Can Play This Game* and *A Lion's Teeth*, and in interviews, conference transcripts, and in personal communications, the metaphysical significance of *Laws of Form* is elaborated in sophisticated and well-defined ways that confirm an interpretation of *Laws of Form* that many have found to be spontaneous and powerfully intuitive. Others and I have appreciated the tremendous power of this work, and in an earlier investigation, I have also brought this metaphysical sense into close connection with what in philosophy is known as first philosophy, in the sense of this term's ancient Greek heritage and its Western metaphysical development.

In "First Philosophy and the First Distinction: Ontology and Phenomenology of *Laws of Form*" (Dible 2023), I described the adventure of the first distinction through the five levels of being as "the first philosophy of the first distinction" (ibid., 520–29), and I have shown how Spencer-Brown's definition of distinction from page one of *Laws of Form* already places the first distinction at the center of the philosophical tradition of the infinite sphere (ibid, 523). There, I also illustrate some of the ways *Laws of Form* can be connected to contemporary phenomenology (ibid, 529–32). In the final section, "The Philosophical Destiny of *Laws of Form*" (ibid, 532–35), I hint at the relevance of *Laws of Form* for Husserl's more general 'infinite task' of scientific first philosophy, and more specific formal ontological task of a "*formal mathesis universalis*" (ibid., 532). In that work, the intention was to lay out the first philosophy of the first distinction, and the purpose, hinted at in the final section, was to build upwards on this foundation. The earlier work on *first philosophy* set the stage for this new work on universal ontology—a kind of *ultimate philosophy*, in that universal ontology is the goal of the ontology of reality, and in that this philosophy of the infinite sphere implied by the paradigm of *Laws of Form* is also a universal ontology of *ultimate reality*.

2 Universal Ontology

What is universal ontology? Universal ontology is, in brief, a speculative metaphysical system. I am employing this designation for the following reasons:

1. It is intended to indicate that the result of a spontaneous interpretation of George Spencer-Brown's *Laws of Form* in metaphysical terms is a universal theory of the being of a distinction within nothingness: a distinction between nothingness and nothingness. More technically, this could be called a universal meontology or a universal theory of non-being, but either way, both amount to the same thing: a way of seeing the transcendental unity, the *Logos*, of being and nothing and everything in between. The unmarked state is spontaneously understood by Spencer-Brown, as well as by any imaginative reader, to be more than a merely formal entity. The unmarked state cannot be said to "be" in the same way that the first distinction can, and neither of these, thought radically distinct, can be said to exist in quite the same way existent things exist: precisely, by dimensional extension.

2. Drawn from the nexus of terminology in contemporary continental philosophy, "universal ontology" has a sense first precisely specified in Edmund Husserl's works but later more fully developed by Husserl's student, Hedwig Conrad-Martius, quite independently of Husserl. The term itself can be found deeper still in early modern philosophy and Romantic philosophy of nature, but I will draw directly on the sense specified by Husserl and fleshed out by Conrad-Martius. Connected to this sense of the term, the ancient tradition of the *mathesis universalis*, fundamental to Husserl's intentions, presents a parallel wellspring of the sense of this term. Conrad-Martius' sense of universal ontology possesses the additional benefit of including in its extant doctrine her recursive philosophical cosmology and physical hyperspace theory. This is the

natural and transcendental doctrine upon which I wish to build this higher synthesis of formal and material ontology.

3. My own sense of "universal ontology" is intended to be a synthesis of Conrad-Martius' metaphysical system and Spencer-Brown's formal-ontological paradigm. My initial metaphysical interpretation of Spencer-Brown's essential elements has remained constant over the years except for the addition of one element: an outermost architectonic element known as the infinite sphere. This addition is also no addition at all since it is implied in the circularity of the first distinction's self-constitutional dynamic ontological closure. The essential interdependence of the center and periphery of the infinite sphere is the first distinction—this latter fact can be seen in Cusanus' naming it at one point "the divine distinction." In my own philosophical biography, the placing of the idea of the first distinction at the center of a metaphysical paradigm (in 2001) represented a first Copernican revolution, and the placing of the infinite sphere around it (in 2015) represented a second. The presentation of this new universal ontology is the result of having undergone a journey through the history of philosophical ideas.

3 The Earlier Universal Ontology: Edmund Husserl and Hedwig Conrad-Martius

Hedwig Conrad-Martius (1888–1966) was a phenomenologist and accomplished philosophical scientist who developed important contributions to the philosophy of essence and existence in the form of a robust ontological phenomenology of nature, ontology of reality, and speculative cosmology placing the natural world within real, actually-extended, higher space-time realms of potentiality accounting for eternity (*aeonische Raum-Zeit*) and infinity (*apeirische Raum-Zeit*). She was also the godmother of Edith Stein, St. Benedicta of the Cross, and was a theosophical mystic of her own unique sort, belonging to the Schobdacher Freundeskreis religious movement. Conrad-Martius came to Göttingen in late 1910 to study with the founder of phenomenology, Edmund Husserl (1859–1938). She attended two of Husserl's winter semester lecture series that have been made available in English as *The Basic Problems of Phenomenology*, by James G. Hart, and *Logic and General Theory of Science*, by Claire Ortiz Hill. James Hart discusses the context of Conrad-Martius' style of philosophizing, called "ontological phenomenology" or "real-ontology," in the preface to *The Basic Problems of Phenomenology*, indicating passages where Husserl develops his own sense of the ontology of reality. It is from such sources in Husserl's thinking that Conrad-Martius picks up the method of, as Hart describes it, a "noematic eidetics or ontology of the real and really real" (Hart 2020, xxiii). In addition to these two lecture series, Conrad-Martius also took courses under Husserl dedicated to Hume, Kant, ethics, and courses under Adolf Reinach, from whom she picked up the emphasis on phenomenological realism.

Around 1913, Husserl's publication of *Ideas* caused a rift in the early phenomenological movement, splitting it into two camps: (a) the idealists who followed Husserl in a kind of transcendental idealism that focuses its energies on the theory of consciousness, and (b) the realists, who represented a more real-world- and object-oriented sense of phenomenology, sometimes identified as ontological phenomenology. Conrad-Martius was the leader the circle of students at that time, and Husserl saw her realist phenomenology as a signal fire for the dissenting side of the debate. Despite the complex historical circumstances of the split between Husserl and the realist phenomenologists, phenomenology continues to develop and differentiate to this day, and many of the diverse developments still call out to be integrated. For our purposes, we hope to see something of the *unity* of real-ontological metaphysics and universal ontology in the work of both Husserl and Conrad-Martius.

In her 1957 book *Das Sein*, Conrad-Martius identifies a litany of kind of things filling out the range of her universal ontology:

> material, formal and categorical objects, ...real and ideal, ...concrete and abstract objects, ...pure entities [qualia], ...purely fictitious objects, ...[and] also purely conceptual objects (40).

We will take a close look at the content of Conrad-Martius' universal ontology after we highlight a few indications of Husserl's sense of universal ontology. In Husserl's 1906/7 lectures, translated by Claire Hill as *Introduction to Logic and Theory of Knowledge* (Husserl 2008b), Husserl gives a similar litany defining his own "universal concept of ontology," which deals with "Being in general in the most universal universality:"

> Being (*Seiend*) in the broadest sense, in that of theory of science as formal ontology, is each and every thing that can figure as the subject of a statement, each and every thing about which we in truth speak, each and every thing that can in truth be referred to as being (*seiend*). And, that does not merely concern things, processes, and what is otherwise real, but even numbers, contradictions, propositions, concepts, theories, ethical, or esthetic ideas, in short the multiform variety of ideal objects, [and] of those that cannot meaningfully be said to have a place in spatiotemporal reality (Husserl 2008b, 97).

In this work—and in many other places—Husserl goes on to distinguish not only formal ontology from material ontology, but also metaphysical ontology, *a priori* ontology, real ontology (the ontology of reality), and universal ontology. Some of these distinctions are well known, for instance, from the tradition of formal ontology that takes off from the *Logical Investigations*, but in some of these works, Husserl only intends to identify parts of the whole program of ontology in order to pursue, for instance, a general noetic eidetics (analytics), leaving the correlative noematic hyletics in the dark. The full circle of (a) noetic and (b) noematic or (a) formal-transcendental and (b) hyletic disciplines, and their general theory appears only infrequently, often in the context of an a priori ontology and universal ontology. There are many other places where Husserl discusses universal ontology, its connection to real-ontology, and its connection to metaphysics.

4 Edmund Husserl's Schematic Outline of Ontology

In Husserl's *Logic and General Theory of Science*, also translated by Claire Hill, in Section 63, "The Ontology of Spiritual Being as an *A Priori* Science of Spirit and Consciousness," Husserl describes an "ontology of spiritual being." This is an ontology of the spiritual layer of such entities as spirit, "ensouled being, the flow of consciousness," and so on—which stand in a realm of nuclear meaning-forms forming an "*a priori* framework" (Husserl 2019b, 298). Conrad-Martius attended these lectures! Earlier in that text, the realm of meaning is described as having an inner structure displaying the amazingly regular pattern of an implemented mathematical theory of forms, resting on "four basic kinds of full-nuclei, ...[forming] fixed configurations into concrete meanings such that all meaning is bound, ... to fixed crystal-configurations and only so crystallized can have concrete being" (ibid., 132). In Husserl's phenomenology, this is the crystallization of the sediment layer that forms beneath the hyletic flux of consciousness. The formal-ontological moment of eidetic analyses in their noetic purity form these crystals in a fixed coherent "crystal system" (ibid., 133). We should bear in mind this formal-ontological framework "resting on four basic kinds of full-nuclei" when we look at the universal-ontological crystal system of Conrad-Martius' framework and also the four simplified expressions of the primitive equations in Spencer-Brown's calculus (condense, confirm, cancel, compensate). The four forms of reference here indicated are Spencer-Brown's fifth canon, called "expansion of reference" in Chapter 3 of *Laws of Form* (Spencer-Brown 2011, 8), and identified, in *A Lion's Teeth*, with the Buddha Sakyamuni's doctrine of *pratitya samutpada*: "What I call expansion of reference, he called conditioned coproduction" (Spencer-Brown 1995, 150).

Husserl opens the third chapter of a later lecture series, published in English as *First Philosophy*, translated by Sebastian Luft and Thane Naberhaus (2019a), by declaring the need for a phenomenological metaphysics, a "metaphysics in a new sense" (ibid., 195), a sense different from the sedimented received tradition that he is often engaged in criticizing. Phenomenologically-reformed and -reconstructed metaphysics achieves some interesting development in *Husserliana 42*, the *Grenzprobleme der Phänomenologie* (2013), not yet translated, in relation to Husserl's own sense of the *prote hyle* and related noematic elements (ideal and real primordial facticity and sensuality). This is clearly a fecund area for further research and development in metaphysics generally, in Husserl's sense of universal science, universal ontology, and theory of science as *mathesis universalis*, but for now, let us see what Conrad-Martius might contribute to phenomenological metaphysics.

5 Hedwig Conrad-Martius' Universal Ontology

Whereas Husserl initiates phenomenological activity through an act called the transcendental phenomenological reduction, in which the phenomenologist actively suspends the positing of hypotheses, including the hypothesis of the world (the world-thesis), in Conrad-Martius' real-ontological reduction, we are instructed to do the opposite operation: the world-thesis is positively enacted, operationalized;

the world is posited as factually real whether or not it factually exists. The real-ontological reduction is the reverse of the transcendental reduction. In the real-ontological reduction, everything is seen as rising forth from its transnatural rootedness in the pre-existing framework (*Voraus gespannten Rahmen*) in which it participates. The real being of the world is hypothetically posited, and things shine forth in their real being and in their self-rootedness, regardless of whether they exist physically, are dreamed, imagined, or hallucinated. Pure ontic research, that is, the ontologically essential exploration of the real world, necessitates this reduction.

In 1913, Husserl published his book *Ideas I*, in which he took his transcendental idealist stance, and that same year, Conrad-Martius wrote her dissertation *On the Epistemological Foundation of Positivism* (*Die erkenntnistheoretischen Grundlagen des Positivismus: Zur Ontologie und Erscheinungslehre der realen Außenwelt*). Like many of Husserl's devoted students at the time, Conrad-Martius took up the mantle of Husserl's demand for an objectivity lying deeper than that sought by scientific positivism. Husserl called for the things themselves, the essence, not just the factual data or datum. Husserl's students were religious converts to the reality that phenomenology opened them up to, but with the publication of Husserl's *Ideas I*, many were sorry to see the master find his certainty in the immanent data of consciousness. Not long after that, Conrad-Martius developed her own method of real-ontology, which was a deepening of the real-ontology belonging to Husserl's earlier period. She spent decades in pure ontic research, developing a view of the pre-existing framework of the physical and the transphysical world through numerous contributions to the special sciences and natural philosophy. It is the totality of this framework and its own self-rootedness in its foundations that constitute her mature universal ontology. The universal ontology of Conrad-Martius is a reflexive cosmology, and cosmogony, in that the whole sphere is tied to the founding ontological dynamism at its center. This kind of universal ontology also absorbs her philosophical cosmology and philosophy of space and time in which ontological interpretations of Einstein's relativity theory (in her 1954 book *Die Zeit*) and Heisenberg's quantum theory (in her 1958 book *Der Raum*) fully develop on the basis of the most classical philosophical definitions of time and eternity, and space and infinity in the Presocratics, Plato, Aristotle, Plotinus, and the Neoplatonists; and the paradoxes of time and motion in Zeno and Kant and modern physics.

Conrad-Martius' ultimate systematic philosophy places the earthly realm between super- and sub-physical spheres (*überphysischen und unterphysischen Sphären*) (Conrad-Martius 1965, 217) that are characterized in a few different ways that will help illustrate Conrad-Martius' universal ontology. The foundation of her system is the movement between being and nothingness that is called "the ontological dynamism," the basic movement (*Urbewegung*) and ultimate root of being (*Seinswurzel*), the leap of being (*Seinssprung*). This dynamism is both a movement from being to nothing and from nothing to being, with the two directions of secondary movement having different characteristics. These two ways of reading the same dynamic are two directions that establish a system of symmetries constituting a cosmogonic polar constitution of all the spheres of being, one of which is the earthly realm. The twofold of the ontological dynamism is characterized in a few ways in her extant works. In one sense, the twofold is on the one hand instatic or infra-static spiritual materiality (*Geiststoff*), and on the other hand an ecstatic spiritual light (*Geistlicht*). In another sense, it is above and below, inner and outer (*absoluten Polen des oben und unten, innen und außen*). In another, itselfness and beingness. In yet another it is center and periphery, one and many, rest and motion, the same and the different. "This is more than a mere analogy" she writes, "it is the recurrence of the same fundamental ontic relationships on an essentially different level of being." ("*Dies ist mehr als eine bloße Analogie; es ist die Wiederkehr gleicher ontischer Grundverhältnisse auf einer wesenanderen Seinsebene.*") (Conrad-Martius 1965, 302–3). All these foundational twofolds are contextualized by foundational three-folds, extensions and reduplications, and ultimately the universal unity of the ontological dynamism grounding them all. In her reformation and implementation of Ancient Greek philosophy, Conrad-Martius' presents a version of Plato's arithmological higher theory of forms, doctrine of the principles of the one (*ta Hen*) and the indefinite dyad (*aoristos Dyas*), in the form of a vision, method, and doctrine of eidetic and ontic scientific philosophy.

6 Conclusion

The ontological dynamism of Conrad-Martius is the material-ontological form of Spencer-Brown's formal-ontological idea of the first distinction. Spencer-Brown's account of creation from the first distinction in the unmarked state could find a sympathetic metaphysical theory-form in the material-ontological dynamics of Conrad-Martius' primordial operation, especially given the primordial expressions of Conrad-Martius' polar system as transcendental causal factors in the metaphysical ontology that results from her real-ontological reduction. What elementary expressions in the Calculus of Indications and its extensions might correspond to the universal justificatory hypostases resulting from Conrad-Martius' real-ontological expansion of reality? Physical reality finds its potential ultimate justification in primordial (first) forces, in Conrad-Martius' universal ontology, and Spencer-Brown's first mathematics appears to accord with the corresponding formal outline, which would be its noetics.

Physical frameworks are too narrow to encompass physical potentiality, and formal categories are too empty to provide for the real physical reality of substantiality and the potential for substantial fullness. A dynamic extension of Spencer-Brown's calculus should look for empirical demonstration in the framework of Conrad-Martius' transcendental physics and universal ontology. Such demonstration could be sought in the conventional conceptions of contemporary particle physics, and Conrad-Martius does indeed provide examples of such demonstration in her ontological interpretations of quantum theory, for instance, but the universal ontological hypostases provide the widest possible indicational space for realization of such formal-ontological dynamics in the substantial fullness that their physical realization demands. Conventional conceptions that are not transcendentally justified according to a primary and central dynamism do not allow realization of form—they are only the guardrails for physical dynamics, emerging from physical necessity and presupposing unjustified hypotheses. What would be the transcendental physical and transphysical real powers that Spencer-Brown would correlate his elementary simple expressions and primitive forms to—in, for instance, his own accounts of creation from the first distinction from nothing, and the five orders of eternity that he identifies? Does Conrad-Martius' universal ontology distinguish levels in the elementary operations following the primary ontological dynamic? Are the laws of calling and crossing present in the emanation and return that describes the morphology of Conrad-Martius' system? How exactly do the realities of the observer and the holocosm as they stand in the recursive cosmologies of both Spencer-Brown and Conrad-Martius relate? Do they share the same universal ontological dynamisms? If so, is this any indication that the system of primitive forms of the Calculus of Indications, or of any dynamical extension of it, might be articulated in a polar organization? Are the symmetries of condensation, confirmation, cancellation, and compensation that characterize Spencer-Brown's Primary Arithmetic reflected in Conrad-Martius' polar organization and fourfold root of elementary natural reality? Finally, could it be that the synthesis of these two transcendental schemas reveal not only mutual corroboration but also mutual deficiencies in their attempts to account for all possible entities? We should follow their examples.

We should not only start from the first principles of analysis, but we should begin by ritually re-enacting in imagination the first distinction from the unmarked state and witness the unfolding of possible forms in the infinite continuum that this ubiquitous act opens. We should dare to imagine in order to see more clearly and to discover the true objectivity hidden by the phenomenal universe. The ideal objectivity of mathematics and the real objectivity of ontology desire one another, and the truth issues from their harmonious encounter. The discoveries of Spencer-Brown and Conrad-Martius are only the first great leaps in universal ontological investigation. Having witnessed their initial developments, we are ourselves initiated and invited to participate in the genius of genesis and universal imagination.

References

[1] Conrad-Martius, H. 1913. *Die erkenntnistheoretischen Grundlagen des Positivismus: Zur Ontologie und Erscheinungslehre der realen Außenwelt.* Halle (Saale): Niemeyer. Diss.

[2] Conrad-Martius, H. 1923. "Realontologie." *Jahrbuch für Philosophie und phänomenologische Forschung* VI, 139–333.

[3] Conrad-Martius, H. 1954. *Die Zeit*. Munich: Kösel-Verlag.

[4] Conrad-Martius, H. 1957. *Der Raum*. Munich: Kösel-Verlag.

[5] Conrad-Martius, H. 1958. *Das Sein*. Munich: Kösel-Verlag.

[6] Conrad-Martius, H. 1965. *Schriften zur Philosophie*, III. Munich: Kösel-Verlag KG.

[7] Conrad-Martius, H. 2024. *Metaphysical Conversations and Phenomenological Essays*. Edited and Translated by Gschwandtner, C. M. Berlin/Boston: De Gruyter.

[8] Dible, R. 2023. "First Philosophy and the First Distinction: Ontology and Phenomenology of *Laws of Form*." In *Laws of Form: A Fiftieth Anniversary*, edited by Kauffman, L. H., Cummins, F., Dible, R., Conrad, L., Ellsbury, G., Crompton, A. and Grote, F. 515—38. Singapore: World Scientific.

[9] Hart, J. G. 2020. *Hedwig Conrad-Martius' Ontological Phenomenology*. Edited by Parker, R. K. B. Cham, Switzerland: Springer Nature.

[10] Husserl, E. 1969. *Formal and Transcendental Logic*. Translated by Cairns, D. The Hague: Martinus Nijhoff.

[11] Husserl, E. 1998. *Ideas Pertaining to Pure Phenomenology and to a Phenomenological Philosophy*. Vol. 1. Translated by Kersten, F. Dordrecht: Kluwer Academic Publishers.

[12] Husserl, E. 2001. *Analyses Concerning Active and Passive Synthesis: Lectures on Transcendental Logic.* Translated by Steinbock, A. Dordrecht: Kluwer Academic Publishers.

[13] Husserl, E. 2006. *The Basic Problems of Phenomenology: From the Lectures, Winter Semester, 1910–1911.* Translated by Farin, I. and Hart, J. G. Dordrecht: Springer.

[14] Husserl, E. 2008a. *Logical Investigations.* Vol. 1. Trans. Findlay, J. N. New York: Routledge.

[15] Husserl, E. 2008b. *Introduction to Logic and Theory of Knowledge, Lectures 1906/07,* Translated by Ortiz Hill, C. Dordrecht: Springer.

[16] Husserl, E. 2019a. *First Philosophy: Lectures 1923/24 and Related Texts from the Manuscripts (1920–1925).* Translated by Luft, S. and Naberhaus, T. M. Dordrecht: Springer.

[17] Husserl, E. 2019b. *Logic and General Theory of Science. Lectures 1917/18 with Supplementary Texts from the First Version of 1910/11.* Translated by Ortiz Hill, C. Cham, Switzerland: Springer Nature.

[18] Spencer-Brown, G. 2011. *Laws of Form.* Leipzig: Bohmeier Verlag. First published 1969 by George Allen & Unwin.

[19] Spencer-Brown, G. [James Keys, pseud.] 1971. *Only Two Can Play This Game.* Cambridge: Cat Books.

[20] Spencer-Brown, G. 1995. *A Lion's Teeth. Löwenzähne.* Leipzig: Bohmeier Verlag.

Practice for Distinction and Context—George Spencer-Brown's Form of Distinction as a Tool for Psychosocial Practice

Tanja Rode
Practice for Distinction and Context, Berlin & Marburg: Germany
`mail@tanja-rode.de`

Abstract

Can an idea as abstract as George Spencer-Brown's form of distinction become directly practical in everyday psychosocial support and guidance? From the perspective of a doctor of political science and counselor, the following article deals with the transfer of epistemological perspectives to the practices of counseling, psychotherapy and forms of reflection;[1] a practice for distinction and context. I explain why, in addition to the form of distinction according to George Spencer-Brown, I need the three pillars of cognition according to Vasubandhu in order to make potentiality a reality. In this sense: I breathe the form of distinction.

I would like to thank Matthias Varga von Kibéd, through whom I am learning new freedoms of thought and grammars of compassion.

[1] In German, we use the term "Supervision."

1 Introduction, or: I Breathe the Form of Distinction

In this text, I unfold my understanding of George Spencer-Brown's form of distinction with the addition of the three pillars of knowledge according to Vasubandhu for my work, my being. It is a personal perspective and a personal train of thought. I consider this to be both theoretically and practically necessary, just as socio-political references, epistemological questions, and fundamental qualities of being must necessarily be affected.

For me, the metaphor of breathing expresses the basality of life and reality.

My professional practice consists of counseling and psychotherapy, forms of reflection, coaching of academic work, moderation and mediation, as well as mediation formats. I work in contexts of social psychiatry and youth welfare, clubs and associations, feminist and queer institutions, inpatient and outpatient support, independent providers and public administration, counseling and therapy, trauma and violence, prevention and intervention. I accompany and support people in professional and personal matters.

My practice is called Practice for Distinction and Context, an explicit reference to George Spencer-Brown. It is to be understood as a name, a designation, as well as an activity and a process.

More fundamental than a specific working format is my epistemological engagement with the philosophical, historical-dialectical materialism[2] and consequent constructivism that I came to know from Matthias Varga von Kibéd. Prof. Dr. Matthias Varga von Kibéd is my most important teacher for systemic thinking and working together with Insa Sparrer, co-founder of the SySt® Institute in Munich. He studied philosophy, logic, and philosophy of science and mathematics at the University of Munich, wrote his doctorate on universal grammar, published numerous writings on systemic work, and has done post-doctoral work on the foundations of formal truth and paradox theory.[3]

---

[2]The nature of its relevance unfolds throughout the text. It would go beyond the scope of this article to even outline it here.

[3]Prof. Dr. Matthias Varga von Kibéd published "Strukturtypen der Logik" (together with W. Stegmüller in 1984) and worked as a professor at universities in Munich, Vienna, Ljubljana, Graz, Constance, Maribor and Tübingen, among

My text is also an unfolding and interrelation of these epistemological perspectives and their practical experiences.

In the following article, I refer to counseling and psychotherapy as the formats of my work.[4] As a professional approach, accompaniment, and support, I see them as part (or an aspect) of life. Counseling is a part of reality, and it creates reality.

I refer to George Spencer-Brown, a British mathematician who lived from 1923 to 2016, and more specifically to his major work, *Laws of Form* of 1969, in which he developed the calculus of indications, an idea that continues to inspire thinking today, for mathematics and science as well as for philosophy. (Varga von Kibéd 2017)

The form of distinction shows that, and how, distinction itself is a complete context. This is outlined in Section 2.

Because the form of distinction operates with an explicit renunciation of truth as a representation of reality (Schönwälder-Kuntze, Wille and Hölscher 2009, 32), I draw on a further source that, for me, represents the reference to reality in a convincing and fundamental epistemological way: the three pillars of knowledge according to Vasubandhu, one or more Buddhist philosophers of the 4th century, as I learned from Matthias Varga von Kibéd. I will explain why and how this is relevant for me in Section 3.

The ideas, as I understand and continue them from George Spencer-Brown and Vasubandhu, are not to be seen as hierarchically built upon each other but as completely interrelated.

In Section 4, I develop these basic ideas for counseling and therapeutic practice. The metaphor of breathing reappears in this sense: for me, George Spencer-Brown's Calculus of Indications is an expression of potentiality and its realization, a basic operation of life, and thus also of my life, the potential of differentiation for every life, understanding, recognition, support. Section 4 explains the extent of the relevance of the encounter with and mutual reference to the form of distinction and the three pillars of Vasubandhu to my idea

---

others. He is currently an associate professor at the Department of Philosophy, Seminar for Philosophy, Logic, and Philosophy of Science at the University of Munich. (https://www.syst.info/de/matthias-varga-von-kibed)

[4]The *differentiation* between the two formats is not the main focus here.

of counseling or therapy, as reality as well as a construction of reality, which in turn can have a changing influence on reality.

I conclude in Section 5 with a summary and an outlook on further potentials of the form of distinction as a form of interpersonal encounter.

2 The Form of Distinction According to George Spencer-Brown

Ever since I was introduced to the idea of distinction by George Spencer-Brown through Matthias Varga von Kibéd, it has been an omnipresent instrument for shaping, understanding, communication; but also even more fundamentally: the most basic form for life itself, (like) breathing, even the form of the creation of the universe.

"Distinction is perfect continence" (Spencer-Brown 1979, 1).

"George Spencer-Brown's idea of distinction is so basic that we don't even know what it is basic for: for mathematics, for logic, for epistemology. It is even more elementary than a system concept, because there is no system in the distinction. Rather, it is about the question of how we produce a difference in the first place, how we begin to construct a system" (Varga von Kibéd 2017, 40ff.).

We, the discerners, only come into being in the process of our discernment. All distinction is also action, not just observation: we point to a clue, we *follow* a task, we produce what is different, we create the difference—and we have reasons, motives for this.

George Spencer-Brown does not say that something *is this way* or that way; rather he asks us to *do* or *see* or *understand* something. That is why *Laws of Form* also begins with the injunction: "*Draw a distinction*" (Spencer-Brown 1979, 3)—and observe what happens to the rest as a result of the distinction. It is about understanding a principle. It is not about concrete distinctions; it is about distinction as such; it is about possibility, about principles of distinction; it is about every distinction.

In this respect, George Spencer-Brown's idea of distinction is fundamental to constructivist systemic thinking. Every distinction is also an act of creation:

respectively I[5] make a distinction; respectively I create one/my world in which this distinction is essential.

Although it is an abstract principle, a potentiality, it is not recognizable without its realization.

Five or six aspects of distinction

There are five necessary and fundamental aspects to every distinction:

- The inner / (de)finite / the definite, that which respectively I point to; the named, that which is at stake.

- The external / in(de)finite / the non-determined, the unnamed, what it is not about.

- The boundary / demarcation (process and result) is both the process of demarcation and the result: the boundary itself.

- The (implicit) context: The context (*the respective one*) is always implicit. As soon as respectively I make it explicit, there is a new designation, thus a new distinction; and there is again a (new) implicit context. It is not possible to make *every conceivable* context explicit. There is always *a* (further) implicit context. Every concrete contextualization is a decision, is again a naming, and is always based on a non-naming. There is always necessarily something that respectively I do not see. Every naming, every inclusion also contains an exclusion. Respectively I can always ask: What is not seen or what is excluded?—but there can never be a definitive answer. It is part of this perspective that knowledge can never be complete. And: I cannot *continuously* ask this question about what is not recognized, what is not included, because then I would *only be* busy questioning sentences, thoughts, perspectives and actions—and would not be able to take any further action.

---

[5] „Respectively I" is a translation (Wolfgang Maiers) of „je ich", already an unusual neologism in German. I borrow "Je ich" from critical psychology (also a form of application of historical-dialectical materialism and itself a "construction") and mean, as there, (respectively) me as an example of people. It could also be someone else. What is relevant, is the explicit reference to subjectivity, to the center of intentionality. I speak of "respectively me," when I mean this exemplary, of "me," when I mean myself concretely and singularly.

- Motive / intentional perspective (aspect of the subject / subjectivity): The named and the unnamed are asymmetrical, of different value due to the motive, due to the (reasons of the, or the) differentiator. The prerequisite for the distinction, for the different value, is initially an inhomogeneity, a difference, which only becomes the basis for the motive by becoming relevant for someone. Drawing a boundary without asymmetry is not a distinction. Without direction, without perspective, there is no distinction (a straight line on a plane, an intersection in space). Only the differentiator with motive, with perspective, with reasons, makes a distinction. The distinctions we make always say something about us, our motives, values, intentions and contexts. About our world, our relative world. Or also: We become who we are due to the nature of our distinctions.

The more complex version of the motive is the:

- Re-entry, for example, the reintroduction of the context into the interior of the distinction. The implicit context can be explicated; it thus becomes part of the named, the new implicit context necessarily becomes a different one—or also: The whole distinction becomes a different one through the explication of the old context. The motive can be reintroduced. Re-entry is, for example, the way in which something essential to someone repeatedly appears in the whole. Re-entry has the character of a process that is modified again and again as it takes place; e.g. thinking about something—without the intention of concluding the thinking, but continuing to reflect on it; always critically questioning what we do; also critically questioning that we critically question. And here, too, it becomes clear that respectively I can do this again and again, but not continuously, because then I would end up in an endless loop. Re-entry can be the aspect of system change.

A distinction is made by creating a border with two sides, so that respectively I cannot get from a point on one side to a point on the other side without crossing the border. On one side (inside) is the named, on the other side (outside) is the unnamed. The named and the unnamed are—via the motive—asymmetrical: of different value.

The five aspects arise in mutual co-dependency or dependent co-generation. They are interdependent. The fact that I speak or write them down *one after the other* is merely due to the language. No single aspect can exist in isolation, as the following questions make clear:

- "How could I have an inside without an outside?
- Or how could I talk about something without something I'm *not* talking about not being thought about, not being heard?
- But how could there be an outside without an inside?
- But how could there be an inside and an outside without a boundary separating the two?
- But how could there be a border if it doesn't demarcate anything?
- But how could there be an inside, an outside and a border—without a space, a context in which this division takes place?
- But how could this space or context exist, unless it is the context *for something*?
- But how could something be the context for something without this something being a distinction?
- But how could something be the context for something that was a distinction if that distinction didn't make a difference *to someone*?" (Varga von Kibéd 2017, 48)

In this last point, the observer/actor, for whom something is at stake, appears as the authority that differentiates. The observer only becomes existent by making a distinction. By placing myself in relation to the world, respectively I *become* the observing, defining and shaping of it—and am observed, defined and shaped by it.

The form of distinction is how we subjectively recognize or move through the world at all. *When* I make a distinction, it is so. When any living being makes a distinction, it is so.

"In mathematics as George Spencer-Brown understands it, no statement is made about what is." Rather, he is concerned with a "reduction process ... of existence to truth, truth to designation, designation to form and form to emptiness." (Schönwälder-Kuntze, Wille and Hölscher 2009, 32)

The idea of existence as something that exists independently is dropped. There is then also no longer any truth if truth is understood as a correspondence between representation and reality. What remains is the idea of reference, which is reduced to the idea of form. Form, too, has no permanence, but must be generated and maintained in ever-new acts; it touches the boundary of disappearance and dissolves when there is no more reference (Schönwälder-Kuntze, Wille and Hölscher 2009, 32f).

"At this stage the universe cannot be distinguished from how we act upon it, and the world may seem like shifting sand beneath our feet" (George Spencer-Brown 1979, p. XXIX).

3 Question: For what do I need Vasubandhu?

Derivation

Thus, every distinction appears as a mere question of decision or agreement, as arbitrary; and even decisions and agreements have no permanence, for we who make them, we who treat the universe, have no permanence either, but exist only in the doing of the distinction that simultaneously co-creates us; we ourselves are sand.

In this understanding of distinction, a sentence can be spoken such as: 'If I only wanted to, I could walk up the walls, I just don't want to right now,' a sentence attributed to George Spencer-Brown himself. This alone, I can call radical constructivism—or nonsense. Not everything is subjective, as long as there is a recognizing, decisive subject:

I have to accept the subject as (objectively, actually) given so that something can happen subjectively, be recognized subjectively, be differentiated subjectively. If I recognize a subject as given in this fundamental way, then also a wall, a force of gravity,

Therefore, I also refer to Vasubandhu in order to explicitly bring the reference to the world into the space (my space of understanding and writing).

That's what I thought in a first draft of the text. In the meantime, I think this is a mistake. The mistake is that George Spencer-Brown is not talking about a real difference that anyone makes.

The form is about potentiality. The real, the actual, is not named, but it is not rejected either. "No statement is made about... Existence" (Schönwälder-Kuntze, Wille and Hölscher 2009, 32) is not synonymous with 'existence is rejected.'

Therefore, my *need* for Vasubandhu is not a necessity in the sense of adding something that George Spencer-Brown has *forgotten*. It is necessary *for me* to apply the form.

Elsewhere, the consideration of the most fundamental principles is about creation: *How God and George Spencer-Brown Created Worlds* is the title of a book by Josef Freystetter in which he examines the analogy between the form of distinction and the biblical creation story (Freystetter 2022).

The beginning of all creation is the transition from potentiality to actuality. A beginning that cannot be understood in terms of time or space. At the same time, potentiality can only be recognized through its actualization.

The moment in which respectively I look at something, create something, is a moment in which reality already exists. Distinctions have already been made. Assuming that something (the world, the real, the interdependent, the living) already exists, I am interested in searching for a form or forms, perspectives or operations that take this idea into account *and* at the same time can be applied as unconditionally as possible, or in any case find application and realization as laws unto themselves. After all, I am not God; creation already exists.[6]

Vasubandhu's three pillars, the triad

The three pillars (dimensions) of knowledge, of constructivism, of being, are

- The interdependent
- The constructed
- The perpetual fact that the one cannot be fully represented/found in the other.

My understanding: there is the existent, everything, in which everything is mutually dependent. There is how we understand, name, grasp, speak: construct this

---

[6] The metaphor works without God and Creator.

interdependence. As a human being, respectively I can never fully recognize the whole, the interdependent. I can only commit, see and grasp that which is accessible, visible and comprehensible to me.

This is not a flaw or deficiency but the natural form of life, of the living. Human constructs (science, models, language ...) never fully grasp the world, the interdependent.

The three pillars are distinct but not separate entities. The separation is theoretical, artificial, and at the same time necessary. But it does not take place on the same level. I can see language, a book, a picture, money both as an aspect of the interdependent (what is) and as an aspect of the constructed (a picture of). Even cognition or perception itself (as an aspect of construction) goes through the whole organism and is therefore also part of the interdependent. Just as I cannot meaningfully ask whether a fabric is red *or* plaid *or* made of silk. In the same way, with the three pillars, I often cannot meaningfully ask: 'Which one is it at the moment?' but rather, I decide which quality I *want* to ask about at the moment.

I also call the three pillars 'triad,' which lets you hear the simultaneity and beauty of a chord.

Reciprocal references by George Spencer Brown and Vasubandhu

In my process, there were different ideas about how I relate George Spencer-Brown to Vasubandhu and how I relate Vasubandhu to George Spencer-Brown.

Because my initial idea was to consider Vasubandhu's pillar of the interdependent as a representation of reality, I *initially* thought of Vasubandhu's triad as a premise for George Spencer-Brown.

But I can also see George Spencer-Brown as the basis for Vasubandhu.

I can also regard the triad as something named in the sense of the first aspect of a distinction. Then it is *part* of a distinction, and therefore not outside and based on it.

I can see each of the three pillars as something named, as the first aspect of a distinction.

I can consider the form of distinction as an aspect of the pillar of the constructed; also as an aspect of the pillar of the interdependent; and even as an aspect of the pillar of the perpetual circumstance of the incomplete visibility of the one in the other, insofar as it is a form, a potential, not a final realization.

I can look at each column broken down by shape.

I can look at the triad as a whole in its form, in each individual aspect, and also as a re-entry.

I can view both perspectives as instruments for being in and observing the world. I can make them reciprocally the object of observation and differentiation by the other.

The triad can also be considered in itself, e.g. as something constructed—and thus also takes on the form of a fractal.

I can also make the form of the distinction the object of itself and thus also have a fractal.

For me—both with George Spencer-Brown and with Vasubandhu—it is about everything: The world, the existing, the given, the interdependent, the living, and humans (and other living beings) as part of it, as cognizers, moving in it, changing, making distinctions, constructing, breathing.

All pillars are shown in the form; the form contains all pillars.

Both can be related to each other in their entirety. One can be fully recognized in the other. Both perspectives can operationalize or deepen each other.

Becoming more concrete, transition to the next section

Respectively I construct, make a distinction; not without presuppositions, but in a world and in relation to a world of interdependence that *is real* and of which I am a part—and which I can never fully grasp. I can see it in such a way that the (form of) distinction is life itself. It is both a reference to the interdependent, a constructive reference, and a realization of or reference to the perpetual fact that the one can never be fully grasped in the other because there is always an implicit context.

For me, the triad is a necessary point of reference in order to understand George Spencer-Brown as a form of the living.

With Vasubandhu, the fact that the interdependent can never be completely found in the constructed is not a problem, not suffering, but natural.

With George Spencer-Brown alone, I experience it as a *problem* because there is no beginning and no end to distinctions. There are distinctions, re-entries, re-contextualizations, new distinctions, *sand*—or the eternally same distinctions as persistence.

This is both a theoretical problem and an actual one: Form-related expressions of human suffering. With Vasubandhu I can formulate this even better as suffering. And by formulating it as suffering, it also contains a possible answer, a path to a solution.

Anyone who has ever tried to fathom any object in depth, anyone who has ever experienced the limits of language, of time, of their own possibilities, knows the feeling that the world melts away between your fingers (or feet) like sand.

And yet, not everything dissolves.

Because, according to my comforting reference, there is the interdependent, of which I am a part and on which I can lean. Distinctions are (also) already there.

I'm not alone—and I'm not the first.

And humans are not even the only and first to make distinctions. All living beings differentiate; distinction *is* life or life *is* distinction. The (form of) distinction is not just cognition, not just language, not just construction, but vitality: It is the form in which I, as a human being, move through the world, through my life, constantly making myself (including breathing, eating, drinking, excreting, making decisions, being professionally active, loving), in constant confrontation and co-creation with the world (which distinguishes, consists of distinctions, changeable as well as less changeable), as part of society, humanity and nature, in conflict about who makes which distinctions and how, so far, now and in the future.

What I can better formulate with Vasubandhu in relation to George Spencer-Brown is: The experience of the melting sand as a subjective problem, suffering, a despair over the perpetual circumstance that the constructed cannot fully grasp the interdependent. Already in the attempt at verbalization, what originally wanted to be expressed changes, transforms, withdraws. In the attempt to formulate, we change who we want to grasp, to verbalize. Historical and social contexts change; people as speaking and showing, listening, and reading counterparts have different contexts and misunderstand each other.

In this sense, I understand the idea that someone could walk up the walls if he only wanted to, with reference to Vasubandhu, as forgetting the interdependent.

With reference to the form of distinction, I can reformulate this forgetting as a confusion of form as possibility with the (form of) actuality; also a reformulation of my own error of thought.

What is important to me:

- that distinctions are processes of life,

- that we do them all the time; they are life itself,

- that we can renew them, question them, make new distinctions,

- that we cannot fundamentally stop doing this,

- but that we have to make a stop every now and then to avoid becoming disoriented or crazy. I can't *keep* asking the question about the unnamed.

- Distinctions are both individual and social processes.

- These are human and generally living processes.

- In this respect, encounter is always also an intersubjective encounter of centers of distinction.

4 Forms of Realization, (My Professional) Practice

Counseling and therapy as a practice of distinction

I understand life as being realized in distinction; counselling and therapy as partial aspects of life practice, as formats of supporting life, as practices of distinction; as encounters between two or more people

- who make distinctions,

- who live in distinctions that others make, that have already been made and will continue to be made,

- who are *fundamentally equal* as human beings in society, the world and nature,

- who struggle for their positions (in the factual as well as in the comprehensible, in the interdependent as well as in the constructed, and in the constant change of the third pillar),

- who have different possibilities, freedoms and constraints,

- who are affected by blows of fate, violence, structures, power relations,

- who are creators and actors.

In this, some are those seeking advice and support; others are those offering guidance and support. Expressing the asymmetry, I understand counseling or therapy as a discerning meta-conversation about different distinctions, as an accompaniment of another person and their distinction(s), offering new, changing distinctions in form and content: Their naming, non-naming, boundaries, setting boundaries, implicit contexts, motives—and re-entries: Naming, retrieval of motives, contexts, thus explications of (implicit) contexts, of world(s) and their perception, thus new distinctions.

Embedded in existing distinctions

Counseling and therapy are not the beginning of distinction, and they are not detached from distinctions that have already been made, continue to be made, and are maintained (through repetition). Counseling and therapy are themselves social practices. They are themselves aspects of social distinctions that they can reproduce (both unknowingly and consciously) or to which they can also relate in a different way.

The prerequisites and framework conditions, and in some cases also the goals of counseling and therapy, are also socially and institutionally structured; relevant distinctions already exist: individual prerequisites set for the person seeking support (such as diagnosis), duration, frequency, treatment goal (health, ability to work). Sometimes the counselor has leeway in which he/she defines his/her own framework conditions. The person seeking advice may have a say within this predefined framework. In other words, the encounter is characterized from the outset by two axes of asymmetry: Who is in need, whose need is at stake, and who determines the framework?[7]

The asymmetry can also be viewed and shaped as: Whose interests are in the foreground, whose life is the benchmark, whose distinctions are centered? However, it is primarily up to the provider of the support format to view and shape the asymmetry in this way. The fact that this is so also has social conditions: If there were sufficient free or low-cost support (measured in terms of need) and therefore more actual freedom of choice, those seeking help would perhaps turn away more often from counseling and therapy contexts whose work is based on claiming interpretive sovereignty over their counterpart, their experience, suffering and solution. This, in turn, is not a personal peculiarity of counselors or therapists, but a component of relationships in which being is divided into wrong and right, sick and healthy.

With regard to other axes of social inequalities, the positions that meet in counseling and therapy can *be* different and be *perceived* differently. Here, visualization, reflection and the possibility of thematization help to avoid the automatisms of

---

[7]This is the difference between privately paid and health insurance-financed therapy, which is free of charge (or financed by health insurance contributions) but involves further access restrictions and framework conditions. Anyone who wants or needs this must be diagnosed and receives a fixed number of hours; and on the other hand, therapy or counseling offer a self-funded client far greater freedom of choice.

social and biographical structures and strategies.

These asymmetries are expressed once again in the language of the form of distinction: I see my practice as distinction-based—and taking place in distinction—supporting others in their distinctions by explicating the questions of who makes, has ever made and can make which distinctions, how and where, including the possibility of bringing in my own distinctions, as a concrete offer and above all as *one possibility* among many. I see the endeavor to refrain from my concrete distinctions and decisions as only one side of the coin (of freedom). The fact that all distinction is a decision and therefore subjective is, in my view, of such fundamental importance that it can and should also show itself in our interaction: The other side of the coin. The clearer, and at the same time, more explicitly reflected, the more freedom.

I support people with their own constructions of reality with regard to their own wishes, perspectives, desires—in their personal paths and feasibility, with the aim of increasing the degree of freedom in the form of becoming flexible, increasing the ability to make decisions with regard to perspectives, sensitivities, perceptions, abilities to act.

Concepts of suffering and trauma

I understand existing trauma terms or concepts, as well as diagnoses, as constructions, also as entity productions of processes and interrelationships. Emphasizing the constructed nature of a term is not about denying the reality of the human suffering that is captured in it. Nor is it about construction *versus* reality. It is not about a specific critique of a concept of trauma that is false or abbreviated—in contrast to *another* concept of trauma that is more complete or reflexive or contextualized.[8]

For me, it is about constant awareness and making it possible to experience as freedom that the interdependent can never fully reveal itself in the constructed; or that every concept is an expression of decisions that can also be made differently.

---

[8] I am pursuing a more comprehensive examination of this in my professional training course "Trauma—more systemic," 2023 in Cologne, at the KIB, an institute for systemic education.

Who makes which distinctions, why and against which background, or which powerful social decisions are made: To develop this together—depending on the needs and concerns of my counterpart—creates new possibilities.

Distinctions that have already been made are factually effective; they are also material violence, they create the phenomenon of the named in the first place (trauma is also socially created). A term not only represents an outline of an object in contrast to what is not named, it also creates realities. The nature of a construction in turn suggests different ways of approaching, different next distinctions, and thus also guides action. By giving trauma the form of a diagnosis, for example, it is presented as an individual illness, treatable and curable.

Just as we cannot do without distinction, we cannot do without language, without concepts. All language, all concepts are in the 'in-between,' come from the 'middle.' There *is* no basis, *no* context. Whatever we do and speak, we are already in the middle of language, in the body, in society, in the world, in existing distinctions. Our movements, including linguistic ones, are also decisions, not arbitrary, but also not under our full control. If we don't forget the possibilities of decision within them, we avoid the reification of the word, of processes, movements, people, life, and give space to the not-yet-seen.

Why a logical form of suffering, trauma, healing ...

Why I find a logical form of suffering, trauma, healing helpful; because through this,

- relaxation can arise,
- all aspects of the event can be discussed, but none of them have to be,
- the client's freedom increases,
- no images and interpretations are fixed,
- a wide variety of experiences can be conveyed,
- individual positioning is possible, but not predetermined, while being made simultaneously invisible,
- different experiences can be broken down without being trapped in interpretations that are already given to everything,

- experiences can be generalized and, based on the generalization, different concretizations can be made possible again,
- experiences can be talked about without activating the emotional shock and catastrophic charge of the term.

Approaches to logical forms of suffering, trauma, healing

I see one of the most general versions of suffering, trauma and healing in the concept of paradox or the system change aspect: Suffering, trauma as a clash of contradictions on a level where they seem to be mutually exclusive, integration or healing on a higher level. Respectively I am then literally no longer the same.

In the language of the triad according to Vasubandhu, I can grasp suffering

- as the absence of one of the pillars (e.g. as loneliness, when pain cannot be communicated; then the pillar of the constructed, of communication, of co-humanity is missing) or as suffering from the discrepancy between the constructed and the interdependent.

I can grasp trauma in the language of form

- as a form of a stuck distinction, as a repetitive, unchanging distinction, without motif, subject, contextualizations appearing, being named and thus being brought in and changed as re-entry,
- in contrast, as a constant questioning of the distinctions made, constant re-contextualizations, new distinctions that let the world slip through your fingers like sand, seem to dissolve your own being,
- in relation to both, as a confusion of the relationship between movement and rest: stuck, or infinite regress, infinite questioning ...

I can grasp healing or liveliness

- as finding or rediscovering a rhythm in the alternation of movement and rest, expressed in the form: in the alternation between standing still with a distinction and setting it in motion again through a re-entry.

I can also describe this rhythm as a moving relationship between remembering and forgetting, and both are process and action, not a thing in the head or the disappearance of the thing. Rather unlearning, learning something else, making a new distinction, dropping another one.

My form of support would be, on the one hand, an invitation to move again within a rigid distinction, starting with explications of various aspects of the distinction found in full recognition of it, and on the other, an invitation to pause, to become aware of the interdependent, (like) a brief pause between inhaling and exhaling.

The re-entry

What constitutes a helpful re-entry depends on the individual and the situation. A re-entry can go both inwards and outwards, bringing the motive and the context into the initial question—and thus broadening perspectives and revealing new options.

A re-entry that is often relevant in my practice is that of contextualizing experiences in larger contexts, in the interdependent, sometimes concretely social relationships: As distinctions that have already been made and that are maintained through constant repetition (or whose repetition, maintenance or change is itself the subject of political debate). The perspective of fractality allows the recognition of analogies. I can also see individual biographies as fractals of social structures.

This recognition of the interdependent often leads to relaxation, affirmation, elimination of loneliness and enables the ability to act on a different level—or contributes to the recognition of what is given and cannot be directly influenced.

In relation to society, I can describe the 'maintenance of power relations' as 'repetition of distinction,' problems as paradoxes, as an aspect of system change, as stuck or endless loop; and social healing as revolution—as radical change, abolition and integration of current contradictions on a higher level.

I understand 'traumatized society,' 'traumatized generation,' or, on the other hand, 'healing is truly revolutionary' (Lucas 2021) as metaphors of this fractality, not as a psychologization of society or a political manifesto for individual healing.

Blind spot and subjectivity

The fact that there is always an *implicit* context cannot be overridden in principle. This contains the humility that I can never fully know the background of my counterpart, that my background of interpretation is not the universally valid one. I practise letting go, unlearning entities and attributions in favour of principles of questions, movement and changing perspectives.

Expressed in the triad of qualities, it is a dance between the interdependent and the constructed (as a change of perspectives) in constant awareness of the third pillar as curiosity and mobility.

In the same breath (a distinction can only be complete), another side of the distinction is the motive, the subject, the respectively I. This is not a mistake to be erased, but a necessary aspect of life, communication, distinction. That is why, for me, an explicit re-entry of subjectivity is an essential aspect of shaping relationships, developing and enabling new perspectives. I am but one, but I am one.

Just as decisions, movements, interactions and processes disappear in formulations such as 'This is so...,' they can also be made visible again in the form of 'I decide it this way' or 'this distinction is often made this way.' These reformulations are the first liquefactions of the frozen relationships from 'this-is-it' to 'this-is-how-it-is-made' or 'this-is-how-it-is-seen' to 'this-is-how-it-can-still-be-seen' or 'this-is-how-it-can-still-be.'

Not a new model but a relationship at eye level

The form of distinction and the triad are not new models but are, rather, offers of differentiation that touch on the question of how models themselves come about and are created, instruments of facilitation. World and self-images, the client's own theories, have complete space in them. My first noble task is to understand these in their own logic, to formulate them as understandable and graspable in principle and concretely, while radically recognizing the subjectivity of distinctions.

These forms are application proposals for searching movements, experiments, questions, aspects of shaping relationships at eye level, instruments of empowerment, democratization, healing.

The higher the level of abstraction (the smaller the individual building blocks), the greater the variety of possibilities for reassembling them. For the greatest possible designibility and multi-perspectivity, the instruments consist not only of building blocks, but also in the processes of building block creation.

The exercise: The form of distinction as an experiential journey & conversation

As part of an advanced training course on "Trauma—more systemic,"[9] which I designed for the first time in this form for the KIB in 2023, I invited participants to take part in the following exercise, with which I wanted to make it possible to experience the form of distinction as an instrument of construction and communication.

Part 1: Guided listening, feeling and thinking journey

Each participant on her own while I speak:

"You are invited to think of a topic of yours, not too difficult, not irrelevant. A matter, a question... —and now try to summarize this in one heading." "Now imagine that you want to explain this topic to someone. What is the topic, what's not the topic?" "How do you draw the line between the named topic and the unnamed topic?" "What is the context against which your topic is relevant for you?" "What do you mean by this distinction? What needs are reflected in this?" "The key words are your 5 aspects of your distinction."

Part 2: Repetition of the five aspects

Opportunity for everyone to write down the five aspects of their own discernment. We come together as a big group: Are there any new insights into the topic as a result of this process? Which ones?

---

[9]Since everything is contextual, there is not only no wrong or right but also no 'systemic.' Matthias Varga von Kibéd therefore prefers to speak of *more systemic*.

Part 3: In small groups

One person names their topic (only this one, i.e. only the first aspect named).
The others discuss on this basis: What could be the unnamed, the boundary, demarcation, an implicit context, the motive; without the topic bringer correcting.
Then a comparison is made: Is the overall context that the group heard and assumed the same as the person telling the story meant?
Experiences: I can only understand 'the topic' if I experience all aspects. I can't draw them from the topic alone. It is interesting to change sequences. Insight: And this is exactly what happens in counseling and therapy, in every encounter, in every conversation.

The greater the asymmetry in the encounter, the more carefully I have to make sure that I have understood the other person's distinctions and not made my own.

5 Summary and Outlook

Aspect summary

I understand therapy and counseling as part of life, as a fractal, as a support for distinction and as distinction itself; understanding, deciding anew; or, related to the idea of paradox, as a (partial aspect of) integration, resolution of contradictions on a new level.

More precisely: Therapy and counseling as support for active reflection on the client's own distinctions, thus contributing to an increase in freedom of decision and action. Personal reflection includes: I am part of the whole, of society, of structures and processes, of the (working) relationship, of the working context. This includes my distinctions, my self-positioning. The more I can reflect on these transparently and name them as *a possibility*, the more space I create for my counterpart to position themselves in relation to them.

Aspect outlook, epistemological and political

The form of distinction according to George Spencer-Brown and the triad according to Vasubandhu together are for me the basis of knowledge and being; here, made explicit for the context of counseling and therapy but equally fruitful for science and politics.

In the form of distinction, I see an instrument, a form of discourse of scientific and political understanding. I can also view scientific standards and content as well as social structures as a multitude of distinctions

- which have been and are being made, and therein name and do not name historically and socially concretely different things (first and second aspect),
- with criteria of distinction, some of which are obviously decided but often formulated as natural and merely 'recognized' (which can be seen on the one hand as a boundary between the named and the unnamed, and on the other hand as pointing to the implicit context; for where motives for criteria of decision do not appear, the framework of distinction also disappears),
- which can be contextualized (context),
- which are decided by subjects who have reasons (motive),
- which are maintained (by reference, repetition),
- which can be addressed (re-entry),
- which can change, which can lead to new, different distinctions (new distinction),
- which can be dropped by letting go of criteria, of boundaries, by unlearning pre-judgments.

On this basis, I can both argue scientifically and see each position as an individual construction. In this respect, this epistemological basis is also a democratic one for me. Attempts to increasingly do justice to the interdependent with the constructed could be made with the aim of not coming closer to 'the truth,' but of grasping and shaping reality in a way that is humane and just to the living, that is just to all, including the recognition that we cannot recognize everything: The possibility of assuming responsibility in not knowing.

References

[1] Freystetter, J. 2022. *Wie Gott und Spencer-Brown Welten erschufen.* Heidelberg: Carl-Auer Verlag.

[2] Rode, T. 2019. Beratung als Praxis der Unterscheidung. Die Unterscheidungsidee von George Spencer Brown, angewandt auf Beratung für LGBTIQ und Geschlecht. In *Körper Beratung*, edited by Bettina Wuttig and Barbara Wolf. 245–272. Bielefeld: transcript Verlag.

[3] Schönwälder-Kuntze, T., Wille, K. and Hölscher, T. 2009. *George Spencer Brown. Lehrbuch. 2.* Revised edition. Wiesbaden: Verlag für Sozialwissenschaften.

[4] Spencer-Brown, G. 1979. *Laws of Form.* New York: Dutton.

[5] Varga von Kibéd, M. 2017. Syst®-Unterscheidungsform: Der elementarste Systembegriff. In *SyStemischer: Zeitschrift für Systemische Strukturaufstellungen.* 11: 40–53.

[6] Lucas, A. 2021. Heilung ist wirklich revolutionär. An interview by Kai Kreutzfeldt. https://annekelucas.com/writing/2021/5/8/interviewed.

[7] https://www.syst.info/de/matthias-varga-von-kibed

POINTING AT NOTHING

FRED CUMMINS
UCD School of Computer Science, University College Dublin
`fred.cummins@ucd.ie`

1 Introduction

Reading *Laws of Form* is not a linear affair. Here, I attend to one of the few footnotes from the text, specifically, to the second footnote to Chapter 8, where Spencer-Brown says:

> Nonmathematicians have attributed some deep philosophical meaning to the unwritten cross, but it is merely a device to collect the whole expression into one bracket. (Spencer-Brown 2011, 35)

As a non-mathematician (in the modern sense), this seems like a promising point of departure. The unwritten cross and the unmarked state are two non-manifest characters that pervade the work. It is difficult to point to nothing.

There is a knowing naivety about axiomatic systems of any sort, for as we begin defining the rules of any mathematical game, we must pretend that we are not really *in media res*, while at the same time, we most assuredly always actually are *in media res*. For mathematics that extends previous work, that builds from known results, this is not a problem. But for *Laws of Form*, that attempts to trace the story "from the beginning," we may perhaps ask what the beginning might mean. We may have an empty page, or whiteboard, or computer screen, but we do not have a nothing to set forth from. As Spencer-Brown/James Keys puts things, "[the excessively 'real' appearance of things] comes through a very clever trick. It depends on an elaborate procedure for forgetting just what it was we did to make it how we find it." (Keys 1972, 31).

Can we dramatize this 'clever trick'? What would it be to 'forget', when what we are asked to forget is everything? There will come an "original act of severance," the drawing of the first distinction, and that act will always be remembered, even if unconsciously, as our first attempt to distinguish "things in a world where, in the first place, the boundaries can be drawn anywhere we please" (Spencer-Brown 2011, xxii).

2   Wonky Geometry

Where are we before anything has happened? This is the Garden of Eden, prelapsarian, ideal.[1] Form is ideal. A circle is ideal. A point is ideal. My incarnate world is wonky, not ideal. Wonkiness is unavoidable. Here is an example:

Drawing a circle

Circles are drawn with compasses. The ideal compass is a secret compass, probably under the care of Arab geometers, made of gold and silver, with a ball at the top to signify the Godhead. One of the two legs approaches the paper orthogonally, and its point must be placed on the paper in a specific place before the second leg is swept around to draw the circle. The placing of the compass must leave a trace, no matter how tiny. Pressure of a metal point upon the fabric of paper deforms the paper. For most of us, this means that there will be hole (probably puncturing the multiple sheets underneath the one we are drawing on) which we will ignore, because we are thinking Form, not Wonky.

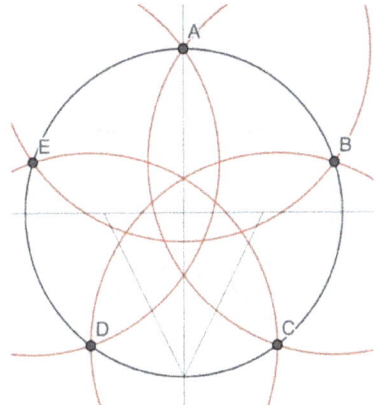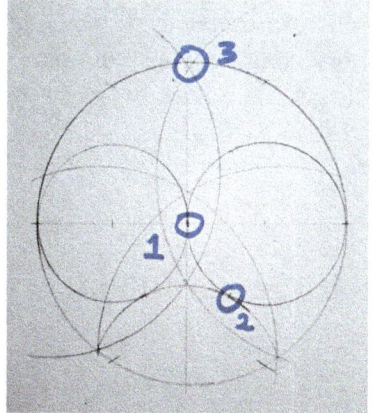

Figure 1: Compass constructions: Left: Desired form as produced on a computer; Right: Wonky construction.

A more elaborate construction

Fig. 1 shows a simple construction with compass and straight line producing a pentagram, or, equivalently, a five-petaled flower within a circle. The construction is simple and well known. It begins with drawing a circle and a diameter. A perpendicular to the diameter is drawn, and two circles of half the original radius are constructed on the two horizontal arms of the diameter. Tracing lines from these centers to the intersection of the perpendicular and the bottom of the circle provides two intersection points. An arc drawn with center at the bottom, and radius provided by these two intersection points produces the intersection points C and D. Arcs with centers at C and D and radius CD produce points E and B, and the same construction centered at either of these two new points produces the top of the flower at A.

This figure was drawn on a computer, using the Geogebra software package (Fahlberg-Stojanovska and Stojanovski 2009). A hand-drawn version is shown on the right hand side. Despite great care, a very small inaccuracy at 1 generates a somewhat larger inaccuracy at 2, which is again magnified in the final drawing of the large arcs, with a resulting manifest inaccuracy at 3. Errors compound. Each step

---

[1]Mythologically, this is the tomb of Christ before the resurrection, just as it is the womb of the Virgin at the annunciation: pure potential.

introduces ever greater inaccuracies. This is wonky, and it looks like a bad place to do elaborate geometric construction. It generates the sneaking suspicion that ever more complex constructions will inevitably bring cascading inaccuracies, and might leave the novice geometer with a sense of futility as thinking in form seems to be restricted to the simplest forms only, while wonkiness has not yet destroyed our work.

Increasing complexity with the Sri Yantra

Figure 2: The Sri Yantra

The Sri Yantra is a contemplative technology from India. Its origins are obscure, and an overview in (Chiodo 2021) takes account of the curiosity this structure has awoken in mathematicians. Chiodo aptly describes the structure thus:

> It can be described from the exterior to the interior as follows. The outermost motive is the square of defence or *bhūpura*. A triple circle circumscribes the core of the diagram. Then, sixteen petals of lotus surround an eight-petalled lotus and a polygonal diagram containing a central point, the *bindu*. (Chiodo 2021, 383).

It is the interior latticework of triangles that concerns us here, and that have attracted geometers from other traditions to inquire into its construction. There are nine triangles: five downward pointing and four upward pointing. Only the largest triangles, one facing up and one down, make contact with the containing ring. The combination of nine triangles establishes a formidable set of concurrency conditions that the yantra must display. These have been communally discovered through repeated development of constructive means for the figure, notably also by Huet (2002). Based on these concurrency conditions, Chiodo presents a construction using straightline and compass (and two parabolas) that spans 340 steps. The entire construction can be visualized here: https://www.geogebra.org/m/cmny6ypg. (We are well aware that this is Greek thinking applied to Indian form. Indian construction methods are not addressed.)

This is an intimidating construction. The considerations of the much simpler constructions with which we began would seem to forcefully suggest that this ought to be undertaken by one highly skilled in the art, capable of divine accuracy, or, perhaps more prosaically, using a computer program.

Wonky construction

And yet form prevails. Fig. 3 shows a Sri Yantra drawn freehand on a whiteboard, beginning with a single circle drawn with a (wonky) whiteboard compass. Simply looking at the sequence of lines on a neighboring screen (the entire inner figure can be drawn as a single line), it was possible, even easy, to approximate things, and to draw the figure in a single stroke. With no measurement or other intervention, simple observation of the model allowed this figure to emerge which satisfies all the concurrency conditions. I find this perplexing and an occasion to consider the relation of form to the incarnate wonky world.

The idealization at the heart of geometry is to be concerned with *space* without *time*. Our wonky world is continuous in spacetime, as far as I can tell. This is what I wish to explore here.

Laws of Form is not obviously a geometric text, yet its initial armory of the cross and associated symbolic equipment is presented on the page, is discussed on whiteboards and the like, and appears thus to be primarily concerned with space. Indeed, its theme is that "a universe comes into being when a *space* is severed or taken apart" (emphasis added). Yet it does not, as Euclid does, begin on the plane.

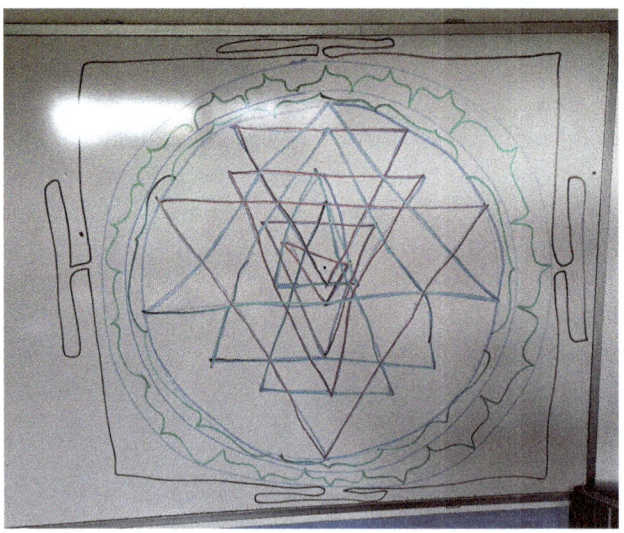

Figure 3: A wonky Sri Yantra, meeting all required concurrency conditions.

Euclid's *Elements* begins with no less than 23 definitions which are required before any postulate (the infamous five axioms) can be considered. The plane is usually understood as a simple concept, but Euclid is particularly vague about what a plane is. A line is a breadthless length (Po. 2); a straight line is a line which "lies evenly with the points on itself" (Po. 4); a plane surface is a surface which "lies evenly with the straight lines on itself" (Po. 7). Despite the obscurity of the notion of "lying evenly" with one's components, I do not remember being confused by the notion of a plane as a child, and I have never seen the idea to invoke confusion in others. The plane is simply taken as the place where the forms live, and a page or blackboard seems to illustrate this notion sufficiently to move on to the serious business of proving theorems.

By contrast, *Laws of Form* begins with very few assumed ideas. The ideas of distinction and indication are introduced and immediately linked ("we cannot make an indication without drawing a distinction"), and the term "form" is introduced elliptically ("We take, therefore, the form of distinction for the form"). The term "drawing" is a conventional verb to use with the notion of distinction, but it seems to necessarily and subtly bring associated ideas of making marks on paper or a surface along with it.

Indeed, the circle in the plane is introduced very early (Page 1) as an exemplary embodiment of the idea of distinction. Space is a unifying notion that runs throughout the main text. Expressions are described as lying in spaces, which is the maximally general notion of a container within which something can lie. Some spaces are deeper than others. Pervasion of one space by another arises. The unmarked state might thus be understood to be also a space, albeit one without a defined limit. This presents no serious conceptual or presentational difficulties until we encounter equations of the second degree in Chapter 11.

Another concept that *Laws of Form* makes abundant use of, without definition, is that of a point. Before the boundary is described as creating distinct sides, spaces, or content, it is necessary to describe the boundary such that "a point on one side cannot reach the other side without crossing the boundary." Much like the plane I encountered in school, this seems entirely simple to grasp, and one notices, only with effort, that another geometric entity has been introduced: the point.

Chapter 11 introduces algebraic expressions of infinite depth, at which point the algebra and the arithmetic become distinct, or "lose contact" with each other. This is where time enters the formal proceedings, and this necessitates the postulation of imaginary logic values, an innovative step in logic, I suspect. Interestingly, in a book of crisp and clear distinctions employing minimal graphical elements, it is also where the limitations of page-based notation become obvious, as the following unnumbered figure demonstrates:

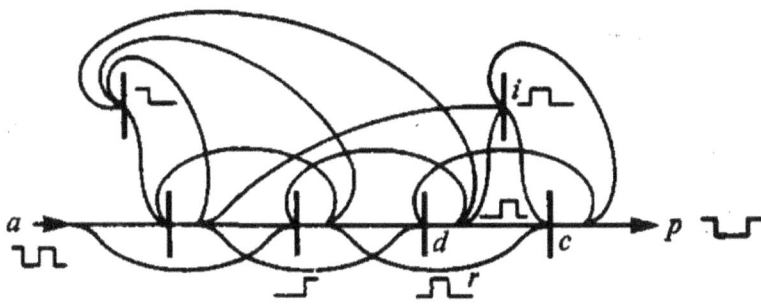

Figure 4: Wonky diagram from *Laws of Form*

As Spencer-Brown notes, "We are now in difficulties through attempting to write in two dimensions what is clearly represented in three. We ought to be writing in three dimensions" (Spencer-Brown 2011, 54). The wonky appearance of the above figure conveys representational difficulties, but the figure works nonetheless. It is formally correct, in much the same manner as the wonky Sri Yantra is formally correct. Mathematicians have overcome worse representational hurdles, as for example, in visualizing the eversion of a 2-sphere in 3-space, which was determined to be formally possible some years before clever graphical artistry made a visualization possible (Reusser 2020).

When we think of space as planar, we may use a page or board on which to write, and because we stand in 3-space, we can look on as outsiders. If we were to write in three dimensions, we would need a volumetric body to inscribe, and we would need to be in 4-space to view our work. I describe my embodied position as lying in 3-space, but that already assumes a rupture of space and time which is an idealization. The notion of "dimension" needs to be briefly addressed. Naive metaphysics assumes time as a line, space as a container, and stuff (*res extensa, cogitans,* or any other stuff that lives there). But that convention refers to how we measure things. In metaphysical discussion, space is 3 because we can readily identify three orthogonal ways in which space extends. This has nothing to do with metricity. The real world can be measured in any number of ways and thus has any number of dimensions. Metaphysics must leave measurement (a thorny problem!) out of things when considering dimensions, and each posited higher dimension is simply orthogonal to the preceding set.

When forms lie on the page, they are clearly manifest in front of us. To the extent that they are seen, any being they might have resides in our capacity to see and interpret what is on the page. Thus they reside in mind, or in the holocosm, to use the terminology Spencer-Brown/Keys introduced in *Only Two Can Play This Game* (1972). Mind is not obviously planar, and form can appear or be produced in mind and then be projected onto whatever is handy in the merocosm.

3 Time and Space

George Spencer-Brown's understanding of time and space is neither to be found in Euclid nor in the Western mathematical canon, and it is not to be conveyed through logical demonstration. In presenting us with *Laws of Form*, he asserts that he is giving us the logic of conditioned co-production (Sanskrit, pratītyasamutpāda, a central term in Buddhist metaphysics). Everything arises, persists, and goes away again, subject to conditions. The details are left as an exercise to the reader.

This reader/writer could do with some help here. Happily, he provides us with a favourable citation in another footnote, this time from *A Lion's Teeth* (1995), where we read:

> The writings of Terence Gray (writing as Wei Wu Wei) offer the best modern independent account of an enlightened view in *Ask the Awakened*, London 1963, and *Posthumous Pieces*, Hong Kong, 1968. (Spencer-Brown 1995, 14).

I encountered the writings of Wei Wu Wei about 20 years ago, and they have never left me. He has many ways of coming at the entanglement of time, space and consciousness, drawing on teachings within the Chan/Zen/Daoist world (roughly speaking). As I have come under the influence of *Laws of Form* (much more recently), the opportunity to think with these two writers together is too good to pass by. There are many affinities between them in their attack on common sense and its pitfalls. Spencer-Brown, for example, regularly stated that there is, and could be, no such thing as a thing. Here is Wei Wu Wei addressing the issue:

> In the absence of the related and interdependent concepts of "space" and of "time" no element of the apparent universe could be conceived, could be cognised or apparently experienced, and no "entity" could be imagined in order to cognise or experience any such element.
>
> Therefore there cannot be any factual entity to be born, to be "lived," or to be "killed," nor any factual object to be brought into existence or taken out of existence. (Wei Wu Wei 2004, 4).

Now let us recall that time and space are tricky notions and that distinction has been elaborated in a purely spatial manner (until Chapter 11). Here, it might be worthwhile to dig into some parallels in the Buddhist literature that seem to draw this out.

Wei Wu Wei troubles our naive understanding of time and space too. He succinctly states: "Time is an imperfect sense of Space. Time is (1) Motion in (2) the Fourth Dimension." (Wei Wu Wei 2003, 6) and he notes that this character of time was made explicit in Kant: "We create Time ourselves, as a function of our receptive apparatus." He also notes that where the Einsteinian revolution in physics changed physics, its simple lesson that time and space are not separable have not been generally learned, though they have profound consequences for our situtated understanding of our own position. I believe this is correct.

Spencer-Brown speaks of a "first" time, that which appears in Chapter 11 as simple alternation. Our clock time he describes as the "third" time, but he leaves few clues as to how this is to be understood. Wei Wu Wei provides a consistent and minimal account that may resonate with this puzzle and encourage us to see Spencer-Brown's cryptic account as drawing from similar intuitions.

As embodied beings, we are in spacetime, which has 4 orthogonal directions (if I follow convention and once more speak of "dimensions", let it be clear that measurement is not being discussed). Geometry is typically held to be concerned with space, but in an ideal atemporal sense. But that does not mean it is about a 2-dimensional plane, or 3-dimensional volumes. Flatland (Abbott 1884) taught us a valuable lesson in thinking of more dimensions than are manifest. In that famous book, folks living in a planar world are visited by spheres which pass through their living spaces generating temporal patterns from which the existence of a third dimension must be inferred.[2]

Wei Wu Wei gives us a way to think of space beyond the first three dimensions. "The fourth-dimension, when seen by us serially in time (as opposed to its total aspect which is eternity) produces the illusion of phenomena" (Wei Wu Wei 2003, 6), and he continues:

> We live in the fourth dimension without perceiving it sensorially, but it is evident everywhere by inference when we know were to look for signs of it.
>
> Duplication, the development of snow-flakes, window-frost, the symmetry of branches of trees, growth of all kinds, radiation, electro-magnetism, motion, light, perhaps undulation, are all probably manifestations of the fourth dimension.[3]
>
> Our psyche exists in the fourth dimension, and our "linga sharira" (composite body which we can only see sectionally). What we see of one another are three-dimensional segments of a four-dimensional totality.

---

[2] It is understood that the planar beings will be enlightened or improved if they can understand the third dimension. Thus mathematicians are presenting us with their own sketch of what enlightenment, or realization, might be, in a minimal form.

[3] Following Ralph Abraham and others, we might add Vibration and Form to this list.

> The next dimension is Eternity (in its time-aspect) and Infinity (in its space-aspect) in which everything exists immutably or is infinite variation at one point.[4] This is the fifth dimension or the second dimension of Time, but Ouspensky states that each higher dimension is infinity for the dimension immediately below it.
>
> The sixth dimension is that in which every possibility exists. (Wei Wu Wei 2003, 7–8)

In *Only Two Can Play This Game* Spencer-Brown speaks of five orders of eternity, and he states that "in eternity there is space but no time" (Keys 1972, 123). This aligns well with Wei Wu Wei's account, where temporal change is understood to be a limitation in our access to that dimension constrained by our receptive apparatus (body). "Three-dimensionality is a function of our senses. Time is the boundary of our senses" (Wei Wu Wei 2003, 7). Elsewhere he says "Quadridimensionally regarded, the Present eternally IS, for Past and Future are not, or, if you prefer, they are Present, but it has no quality of time, that is no kind of duration or continuity (succession)" (Wei Wu Wei 2003, 81). This would be a geometric (hence, ideal and timeless) way to regard the four dimensions.

When Spencer-Brown and Wei Wu Wei both speak of the first four dimensions, they are encouraging us to do something similar—to conceive of our situated finitude and to go beyond that, by making time atemporal (though manifestation to an embodied being is clearly temporal). Both authors/sages have constructed a way to think beyond the three of space to the four of spacetime. But they both extend the process and do to those four dimensions what they did to the three familiar spatial ones. The fifth will not be easy to speak of. Spencer-Brown says:

> "Everything reflects in everything else, and the peculiar and fundamental property of the fifth order of being reflects itself all over the universe, both at physical and metaphysical levels. An interesting reflexion of it in mathematics is in the fact that equations up to and including the fourth degree can be solved with algebraic formulae. Beyond this a runaway condition takes over making it impossible to produce a formula to solve equations of the fifth or higher degrees. (Keys 1972, 127/128).

---

[4]The astute reader will note that this is a simultaneous consideration of the magnificent but normally irreconcilable visions of Parmenides (immutability) and Heraclitus (variation). It thus swallows all being and becoming whole.

Interestingly, Wei Wu Wei spoke of the fifth dimension as the second dimension of Time. Neither makes confident claims about characterizing a sixth or subsequent dimension: "In the deepest order of eternity there is no space. It is devoid of any quality whatsoever" (Keys 1972, 123). "The sixth dimension is that in which every possibility exists" (Wei Wu Wei 2003, 8). Wei Wu Wei helpfully adds:

> The notion that there are only three dimensions is primitive. In fact we only know how to make use of three. In any case a dimension is not a thing-in-itself: it is an intellectual instrument. There are as many as we care to use, as many as we may need. The fourth exists neither more nor less than the second. Their purpose is to help us understand the phenomenal universe which surrounds us and of which we are a part. As long as we limit them to three we are able to understand at most the physical part of our being and the outside of everything that is accessible to our senses. A dimension can never be anything but a mathematical concept. (Wei Wu Wei 2003, 30/31)

4 Discussion

The unmarked state has all possible potential, as does the Garden of Eden, until something happens, and we fall into the wonky world. The unwritten cross is the gentlest possible way to point to this. Philosophy is a fine sport, but this kind of non-dual seeing[5] is not discursive and is not to be explicated propositionally. This, I take it, is why Spencer-Brown chooses not to attach any "philosophical significance" to the notion.

My perplexity at drawing the complex triangles of the Sri Yantra remains. But I am reminded at this point of the manner in which cat's cradle games function in many traditional cultures. Formal patternings of intricate design are created out of looped pieces of string. Children play such games, and they are also an important way in which formal patterns are preserved and transmitted in the wonky world. An illustrative video recording from 1911 of children of the Amazonian Pemon or Taulipang people in Guyana playing such games illustrates the point well

---

[5]I understand the slightly odd way Spencer-Brown talks about seeing but not with human eyes to be yet another attempt to convey what is otherwise often called *non-dual seeing*, e.g. in the tradition of Advaita Vedanta.

(Koch-Grünberg 1911). In such games, complex topological figures attract symbolic names (e.g. Jaguar's Mouth) and enter the common stock of formal knowledge (Fig. 5). Compass constructions may be a refinement of such formal embodied practices. The Yantras too. I don't really know.

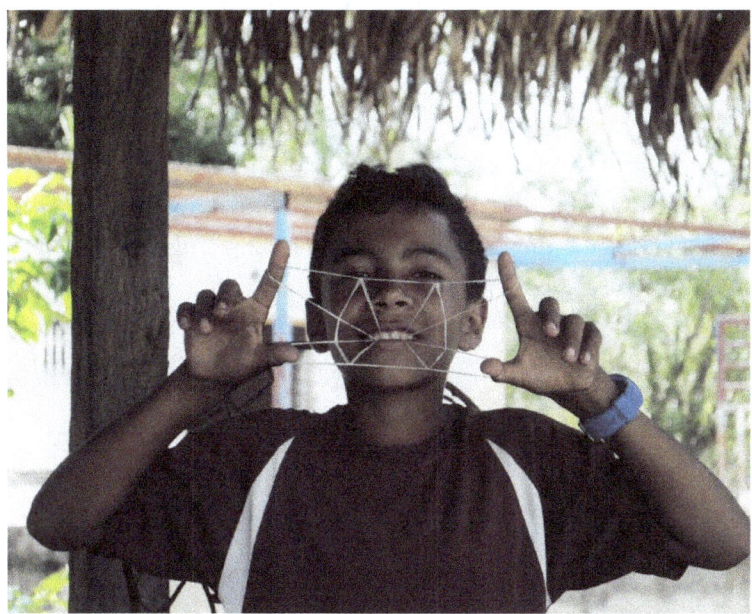

Figure 5: A Jaguar's Mouth pattern by a child of the Yukpa people (Columbia). Other patterns include figures of 15, 18 and 20 vertices, though I have not provided images of these patterns. Examples may be found at https://www.alysion.org/figures/kindofeasy.htm.

References

[1] Abbott, E. A. 1884. *Flatland: A Romance of Many Dimensions.* Seeley & Co., London.

[2] Chiodo, A. 2021. On the construction of the śrı yantra. *Comptes Rendus. Mathématique,* 359(4):377–397.

[3] Fahlberg-Stojanovska, L. and Stojanovski, V. 2009. GeoGebra—freedom to explore and learn. *Teaching Mathematics and its Applications: An International Journal of the IMA,* 28(2): 69–76.

[4] Huet, G. 2002. Śrı yantra geometry. *Theoretical Computer Science*, 281(1-2): 609–628.

[5] Koch-Grünberg 1911. Koch-Grunberg—Aus dem Leben der Taulipang in Guayana—Filmdokumente aus dem Jahre 1911. Video recording of cats cradle games among Amazonian natives. URL = https://youtu.be/frvlMiHxvQ0?si=GR1yHgDwxIKwhDQS&t=386.

[6] Reusser, R. 2020. Sphere eversion. URL=https://rreusser.github.io/explorations/sphere-eversion/.

[7] Spencer-Brown, G. 2011. *Laws of Form*. Bohmeier Verlag.

[8] Spencer-Brown, G. 1995. *A Lion's Teeth*. Bohmeier Verlag.

[9] Spencer-Brown, G. [James Keyes, pseud.]. 1972. *Only Two Can Play This Game*. New York: Julian Press

[10] Wei Wu Wei 2003. *Fingers Pointing at the Moon*. Sentient Publications.

[11] Wei Wu Wei 2004. *Posthumous Pieces*. Sentient Publications.

Laws of Form in Kindergarten

Moshe Klein

mosheklein@mail.tau.ac.il

Abstract

Through a dialogical approach in which an adult explores problems with children, it is possible to teach new concepts in mathematics. It is possible to teach children in kindergarten Spencer-Brown's *Laws of Form* and thereby fulfill Leibniz's vision.

1 Introduction

The mathematician and philosopher Gottfried Wilhelm Leibniz (1646—1716) wanted, as a youth, to develop a universal language with only one symbol (Dascal 2008). He believed that discovering that language would help settle disputes between people.

Our research is based on work done in kindergartens under the topic of developing a new mathematical language. Children love calculating and do it naturally and spontaneously before they are actually taught arithmetic in an orderly and systematic fashion.

From Piaget's research on the development of children's thinking, mathematical principles of conservation and symmetry were extended to the psychological field (Piaget 1923). Later, Vygotsky showed how important interaction between child and adult was, where the child could understand and succeed much faster in the presence and under the influence of an adult (Vygotsky 1987).

Another point to make here is Reuven Feuerstein's research that stressed the child's learning and development abilities distributed over time, rather than their capabilities being fixed at a given moment in time (Feuerstein et al. 1991).

Passing on knowledge, which was the traditional and classical role of the educator, has changed, since children nowadays are themselves highly exposed to information sources over the internet, computers and various applications.

The educational approach suitable for today is that of **the pedagogy of the unknown**. The basic idea behind it is that the kindergarten teacher learns together with the children.

The pedagogy of the unknown creates a clear distinction between two concepts: presentation and experiment. In a presentation, the results are known to the adult performing them in front of the children. In an experiment, the adult is himself curious about the unknown results. In this case, when the adult guides the children in the process of investigation, a research community is created, where neither adult nor children know of the outcome.

2 Research Conducted in Kindergartens

Research results are based on the instruction of 120 kindergarten teachers in the implementation of the system for mathematical topics. We offered them mathematical activities with their school children where they themselves had to investigate and develop together with the children. These topics aroused great enthusiasm and interest with both children and their teachers.

We observed, in this process, that kindergarten children are capable of dealing with deep abstractions of the concept of number.

The mathematical topics we chose to focus on were:

a. **Permutation**.
Permutation is the placing of objects in a row. For example, 6 permutations of three objects:
$$123, 132, 213, 231, 312, 321$$
Kindergarten children love and succeed in discovering the 6 permutations of 3 and succeed in finding the 24 permutations of 4.

Figure 1: Children investigating the permutations of 4

b. **Number partitions**
Number partitions are the various ways a sum of smaller or equal numbers is presented. For example, the 3 partitions of 3:
$$3 = 1 + 1 + 1$$
$$3 = 1 + 2$$
$$3 = 3$$
Kindergarten children love to find number partitions, which they do with relative ease with $4, 5, 6$. With regard to the way in which the equation is presented, it is important to note the difference between expressing it as

$2 + 3 = 5$, which is an arithmetic exercise with only one solution as opposed to: $5 = 2 + 3$, which is one possible partition out of 7 different ones. When the children are busy with number partitions, they uncover the open, creative and playful nature of mathematics.

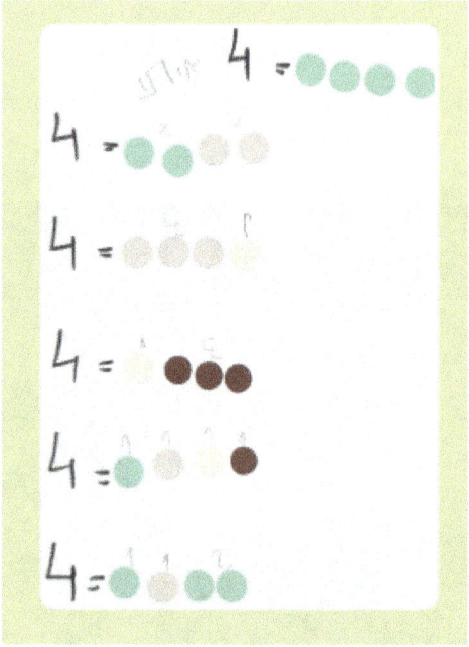

Figure 2: A child discovering the 5 partitions of 4

c. **Number forms**

Number forms, as a concept, were developed and inspired by the book *Laws of Form* by George Spencer-Brown (1969), which is a generalized plane of number partitions. We investigate the relations between circles that do not intersect. Number 3 would have 4 different forms as shown in Figure 3:

Number forms can be illustrated by playing with bamboo circles of various radiuses. You present the forms of 3 and then the children manage to investigate 4 in an abstracted way with the use of paper and pencil (Fig. 4).

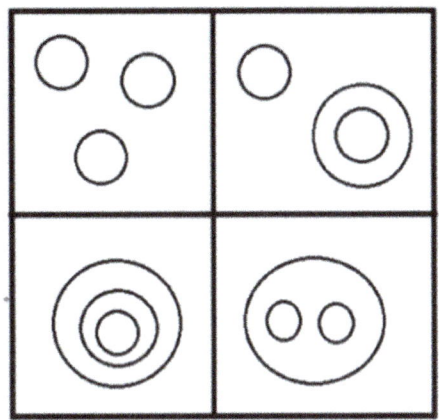

Figure 3: Four forms of Number 3

Figure 4: A girl describing the 9 forms of 4

The following is a table summarizing investigations conducted by kindergarten children:

forms	partitions	permutations	number
1	1	1	1
2	2	2	2
4	3	6	3
9	5	24	4
−	7	−	5
−	11	−	6

3 Conclusions

Kindergarten children love mathematics and are highly curious about its topics within their reach. Their advantage lies in them being free of hidden assumptions that we adults have, for example, the notion that a line is made up of dots.

Working on mathematics with kindergarten children through a sincere dialogue enables us to discover new thinking forms on our own, like Ramanujan, who discovered 4000 mathematical equations without being able to explain why they were correct.

Thanks to working in the vivid space of kindergartens, we developed a new number which is soft number (Klein and Maimon 2021) which is based on the distinction between the multiplications of zero. Discerning the various forms of 0 can serve as the basis for the development of a mathematical language with a soft logic, which was one of Leibniz's ambitions.

References

[1] Dascal, M. 2008. *Leibniz: What Kind of Rationalist?* Logic, Epistemology, and the Unity of Science, 13. Tel Aviv: Springer.

[2] Feuerstein, R., Klein, P. S., and Tannenbaum, A. J. 1991. *Mediated Learning Experience (MLE): Theoretical, Psychosocial and Learning Implications.* Freund Publishing House Ltd.

[3] Klein, M. and Maimon, O., 2021. "Fundamentals of Soft Logic," *New Mathematics and Natural Computation*, 17(3): 703-737, 2021 http://dx.doi.org/10.1142/s1793005721500356

[4] Piaget, J. 1923. *The Language and Thought of the Child*. London: Routledge.

[5] Spencer-Brown, G. 1969. *Laws of Form*. George Allen & Unwin.

[6] Vygotsky, L. S. 1987. *The Collected Works of L.S. Vygotsky: The Fundamentals of Defectology*. Vol. 2. Springer Science & Business Media.

The Spencer-Brown Society Constitution

1 Name

The name of the group shall be **The Spencer-Brown Society** (the Society).

2 Aim and Powers

a. The aim of the Society is to promote and advance the work of Professor George Spencer-Brown, in particular by:

   i. organising and conducting conferences, lectures, and other events, nationally and internationally;

   ii publishing books, journals, and other material, in print and in digital media;

   iii managing and disseminating information about Spencer-Brown's work and legacy;

   iv encouraging scholarship and research.

b. In furtherance of the Society's aim, the Committee has the power to:

   i. raise funds, receive grants and donations;

   ii. buy or sell property, take on leases and employ staff;

   iii. appoint trustees to hold property;

   iv. cooperate with and support other groups with similar purposes;

   v. do anything else within the law which is necessary to achieve the aim.

3 Membership

a. Membership is open to all who are in accord with the aim of the Society.

b. The Society shall not discriminate on grounds of gender, race, colour, ethnicity, nationality, sexuality, disability, religion, political belief, marital status, or age.

c. Members are required to treat other members with respect at all times and to conduct themselves with goodwill and in a convivial manner.

d. Members may vote at General Meetings of the Society.

e. The Society shall maintain a list of members.

f. Members are required to keep the Secretary informed of their current email address.

g. Members may be required to pay a subscription at the discretion of the Committee, to be confirmed by the members at a General Meeting.

4 Membership Termination

a. Members may resign at any time by emailing the Secretary accordingly.

b. Membership may be terminated if any subscription owed to the Society remains unpaid three months after payment became due.

c. The Committee may expel a member for conduct prejudicial to the Society, provided that any member whose expulsion is proposed shall have the right to be heard by the Committee before a decision is made. Anyone expelled from membership of the Society may not subsequently be re-admitted to membership without special approval by the committee.

5 Officers and Committee Members

a. The business of the Society shall be conducted by a Committee elected at the Annual General Meeting (AGM).

The Spencer-Brown Society Constitution

b. The Committee shall consist of three Officers. The Officers may, at their discretion, co-opt onto the Committee no more than two additional Committee members.

c. The Officers' roles are:

   i. Chair, who shall chair General and Committee meetings;
   ii. Secretary, who shall maintain the member list, take minutes, and distribute communications;
   iii. Treasurer who shall maintain accounting records.

d. In the event of a vacancy occurring in any officer post, the Committee may appoint someone to fill that post until the next General Meeting.

e. The role of Committee members, who are not Officers, shall be to manage the membership, events, publicity, marketing, publications, and other business of the Society.

f. Any Committee member not attending four consecutive Committee meetings without apology shall be deemed to have stood down.

6 Annual General Meetings

a. An Annual General Meeting (AGM) shall be held at least once in each calendar year, and no more than 15 months after the last AGM.

b. AGMs are open to all members of the Society.

c. The Secretary shall email members with the date, time, and venue, not less than fourteen days before the AGM.

d. AGMs may be held online provided that each participant may communicate with all the other participants.

e. Nominations for election to the Committee shall be made to the Secretary in such manner as the Committee may from time to time direct.

f. The quorum for the AGM shall be ten percent of the membership.

g. At the AGM:

  i. The Committee shall present a report of the Society's work during the previous year.

  ii. The Committee shall present the Society's accounts for the previous year.

  iii. The Committee shall stand down.

  iv. The Committee for the coming year shall be elected.

  v. Matters communicated in writing to the Secretary not less than seven days ahead of the AGM shall be discussed.

7 General Meetings

a. A General Meeting (GM) can be requested by a majority of the Committee or by ten percent of the membership by making a written request to the Secretary.

b. General Meetings are open to all members of the Society.

c. The Secretary shall set a date for the General Meeting which shall be held within twenty-eight days of the Secretary receiving the request.

d. The Secretary shall give members not less than fourteen days' notice of a General Meeting, stating the venue, date, time, and agenda.

e. General Meetings may be held online.

f. The quorum for a General Meeting is ten percent of the membership.

8 Committee Meetings

a. Committee meetings may be called by the Chair or Secretary. Committee members must be given notice of Committee meetings not less than five days before the meeting.

b. The Committee shall meet as it determines necessary.

c. The quorum for Committee meetings is three Committee members, of which two must be officers.

　　d. Committee meetings may be held online.

　　e. Members who are not Committee members may attend Committee meetings at the invitation of the Committee. Members who are not Committee members may not vote at Committee meetings.

9 Rules of Procedure for Meetings

　　a. The Chair of the Committee shall preside at all meetings but if he or she is not present, those present shall choose one of their number to chair the meeting.

　　b. Matters arising at any meeting shall be discussed openly and the meeting shall seek to establish a consensus broadly acceptable to members attending the meeting.

　　c. If such a consensus cannot be achieved, a vote shall be taken, and a decision shall be made by a simple majority of members present. If a vote is tied, the matter must be taken to the members in a General Meeting for resolution.

10 Property

　　a. All or any part of the property of the Society may be vested in not less than two Holding Trustees (or in a corporation entitled to act as custodian trustee) appointed by the Committee, and such Holding Trustees shall hold such property and deal with it in a manner which is consistent with the aim of the Society as the Committee may from time to time direct. The powers, rights and duties of Holding Trustees so appointed shall be embodied in a Trust Deed to be approved by the Committee and to be executed by the Holding Trustees. Provided they act only in accordance with the lawful directions of the Committee, Holding Trustees shall not be liable for the acts and defaults of its members.

　　b. The Committee may at any time remove or replace any Holding Trustee and may appoint a Holding Trustee in place of any Holding Trustee who has retired, dies, refuses to act or has become incapable of acting.

c. Any property or contracts, including contracts of employment, held in the name of the Society and not vested in named Holding Trustees shall be deemed to be held jointly by the members of the Committee for the time being.

11 Finances

a. A bank account shall be maintained on behalf of the Society at a clearing bank determined by the Committee.

b. Three cheque signatories shall be nominated by the Committee, one of whom shall be the Treasurer.

c. All payments shall be authorised by two signatories, one of whom must be the Treasurer.

  i. For cheque payments, the signatories shall sign the cheque.

  ii. For other payments (including BACS, cash withdrawals, debit card or cash payments), written authorization shall be provided by two signatories and retained by the Treasurer. Written authorization may be provided by email.

d. Records of income and expenditure shall be maintained by the Treasurer and a financial statement given at the AGM and on request at any meeting.

12 Not-for-Profit Status

a. No part of the income or property of the Society shall be transferred, directly or indirectly, by way of dividend, bonus, or otherwise by way of profit to any member or Committee member of the Society, provided that nothing shall prevent any payment in good faith by the Society:

  i. as repayment of reasonable out-of-pocket expenses incurred by any Committee member whilst acting on behalf of the Society;

  ii. of interest on money lent by any Committee member at a rate per annum not exceeding 2 per cent above the base lending rate of the Society's bankers for the time being;

iii. of reasonable and proper rent for premises demised or let by any Committee member of the Society.

13 Changes to this Constitution

a. Changes to this constitution may only be made at an Annual General Meeting or a General Meeting.

b. Any proposed change to this constitution must be made in writing to the Secretary. The Secretary shall then email the members with the proposal and notice of the meeting.

c. Any proposal to change the constitution shall require a simple majority of members present and voting at the meeting.

14 Dissolution of the Society

a. If twenty percent of members express a wish to dissolve the Society, they may call a General Meeting by writing to the Secretary. The sole business of such a General Meeting shall be to determine, by simple majority vote, whether the Society should be dissolved, and, if so, to dissolve the Society.

b. Upon a simple majority vote to dissolve the Society, the Society's funds, after all outstanding debts have been discharged, shall be donated to a registered charity. The recipient charity shall be decided by simple majority at the General Meeting in which dissolution occurs.

c. If the Society is dissolved, any assets remaining after the satisfaction of its debts and liabilities shall not be distributed amongst the members but must be transferred to one or more non-profit or charitable organizations with aims similar to, or compatible with, those of the Society.

www.ingramcontent.com/pod-product-compliance
Lightning Source LLC
Chambersburg PA
CBHW082039230426
43670CB00016B/2715